AMPHIBIOUS
SOUL

AMPHIBIOUS SOUL

FINDING THE WILD IN A TAME WORLD

CRAIG FOSTER

HarperOne
An Imprint of HarperCollinsPublishers

FIRST EDITION

Designed by Janet Evans-Scanlon
Track illustrations by Ian Thomas and Alex van den Heever of the Tracker Manual

Tracking Journal:
Page texture: © Olex Runda/stock.adobe.com, © Little River/stock .adobe.com
Photo texture: © macondos/stock.adobe.com
Photo shadows: © elena_garder/stock.adobe.com
Arrows: © samemen/stock.adobe.com

Library of Congress Cataloging-in-Publication Data has been applied for.

ISBN 978-0-06-328902-4

24 25 26 27 28 LBC 5 4 3 2 1

I dedicate this book to the natural world for being my greatest mentor, guide, and inspiration, and to our great ancestors who ushered us through so many difficult times in prehistory.

CONTENTS

INTRODUCTION

IN SEARCH
OF WILDNESS

THE PANHANDLE IN THE OKAVANGO DELTA IN NORTHERN BOTSWANA is a primal place where people still have dangerous encounters with big animals like hippos and elephants. The air is thick with buzzing insects and birdsong, and the great flowing river is a giant silver serpent fringed with vast beds of papyrus reeds. Most rivers run to the ocean, but the Okavango flows into this great swamp—it is the life giver.

There is an intensity here that wakes me up.

The swamp is home to an incredible diversity of animals who play a role in this rich ecosystem, from the sitatunga antelope, whose banana-shaped hooves allow them to move silently through the marsh, to the hippo, who carve pathways through the dense aquatic grasses, directing the flow of water and making life in the grasslands possible for so many.

But the animal our small filmmaking crew was searching for this golden afternoon was the Nile crocodile, a great prehistoric creature that is the largest freshwater predator in Africa.

As our motorboat skimmed along the channel's fringe, we trained our eyes and cameras on the banks where the crocodiles commonly bask in the sun. Here the papyrus grows thick, its tall grass fanning out into bright green rays that resemble a bursting firework. Below the swaying papyrus culms are networks of tight tunnels and dark caves, the underwater dens where the crocodiles drag their kills.

Naturalist and local guide Greg Thompson led our small crew, which included my brother and longtime filmmaking partner Damon; the French underwater cinematographer Didier Noirot, who had worked on Jacques Cousteau's team; and my friend Roger Horrocks, one of the best underwater cinematographers in the world and as close to me as another brother.

Roger was the reason I was here—why our team was about to dive with the species considered to be among the most dangerous on the planet. He and I had met four years before, at a film festival in Durban, South Africa, and we clicked immediately. Roger is a deep thinker, a philosopher, and a man of action. He's also one of the most accomplished divers I know, entirely at home in the water. If Roger thought we could follow the crocodile into its lair, I was interested in seeing if it was possible. We knew from diving with great white sharks that big predators are not always as dangerous as Hollywood movies make them out to be.

As Greg guided our team closer to the dragon's lair, he shared a warning. An experienced adventure guide who gives tours out of a wood-paneled two-decker houseboat called the *Kubu Queen*, Greg knows the Okavango Delta panhandle well, and he knows crocodiles.

He couldn't guarantee we would survive the dive.

Crocodiles are among the few predators that consider human beings prey, and they have the strongest bite of any animal. Lying in wait at the water's edge, this ambush predator will attack and eat almost anything that comes into the water—and every animal must come to drink eventually.

"Humans are the perfect prey for these crocodiles," Greg pointed out. "We're the perfect size."

As our boat motored slowly along the curving stretch of water, Didier skimmed the surface with a mini high-definition camera on a long pole that resembled a pool cleaner.

After a few moments of filming, Didier called us close to view a playback of his footage.

"I've never seen this before," he said, half joking. "Maybe he was ambushing us."

I looked over the side of the boat to track what he had captured on film—and there was a crocodile some fourteen feet long. He was extraordinary, an ancient and graceful dragon more than twice as long as a man. Though the crocodile was swimming close to the boat, there was nothing aggressive about his behavior, and we began to prepare ourselves for the dive to get a closer look.

Before anyone broke the surface of the water, we checked for hippos. Despite its rotund appearance, the hippo is the most dangerous animal in the Delta. These massive beasts, weighing up to four tons, with teeth a foot long, can move rapidly underwater, running along the bottom of a riverbed. A hippo can even lift and capsize a boat. We had met people who had witnessed hippo attacks in which someone had been maimed or killed. If a hippo was near, we would have to get out of the water immediately.

Roger and Didier slipped in, followed by Damon. They moved quickly, and as quietly as possible to avoid alerting the croc with a big splash. The surface is the riskiest place in the water. It is where the crocodiles prefer to attack their prey and wrestle it down to the depths. So the divers had to immediately descend and stay at the river bottom, then come straight back up and into the boat without lingering at the surface.

I stayed above to film as the three descended to where the crocodile was now resting on the sun-dappled sediment floor. They avoided moving backward as a prey animal would; instead, they kept still or moved toward the croc, a tactic to confuse the animal.

I started preparing my gear to follow them into the water. I was younger then, more foolish and willing to take risks, but still I was scared. These animals are territorial. A member of a photographic expedition that traveled to this same place not long after our own expedition had his arm torn off by a crocodile and nearly died. Still, despite the risks, I felt the pull to be closer, to feel this animal's wildness, to understand.

Of course, we had taken precautions. For most of the year, even the most daring diver would not attempt to swim the waters of the Okavango Delta—visibility is too low, leaving you with little way of knowing if a crocodile is approaching. But we'd planned our trip in June, when the surge flows peak—the water flows strongly and steadily, clearing the sediment, and there is usually a two- to four-week window of clear water.

Waiting on standby in the boat was a trained medic with a fully stocked trauma kit, including oxygen cylinders. When the dive team slipped beneath the surface, the medic went on high alert. We had briefly considered taking some kind of weapon with us but decided it would be unfair to invade an animal's environment and then try to kill it if it attacked. Also, if a crocodile was going to attack us, we probably wouldn't even see it coming. This is such a large and powerful predator: there was no weapon that was going to help us.

After Roger, Didier, and Damon had spent about forty-five minutes filming the croc on the river bottom, the dragon slowly glided to the surface for a breath and then moved almost lazily toward the papyrus, stirring puffs of sediment as he went. Lungs filled, he dove back under—and then something extraordinary happened. As I watched from above, the crocodile began to walk deeper under the cover of the papyrus. When I think about it, I still get gooseflesh.

It felt like an invitation to follow. For weeks, we'd been tracking these crocodiles in the hope one would lead us to its lair, but to no avail. But now it seemed this one was doing just that. Surely here was our guide, and with him, perhaps, the wildness that had so long eluded me.

I pulled my mask on and tested my regulator. Just before going in, I noticed a strange sight: a bat, flying around the boat in broad daylight. I thought, *Something powerful is about to happen.*

I went over the side and found myself in a parallel world, an underwater garden tinted emerald green and gold from the sunlight filtering through the water.

With the help of camera lights that broke through the darkness and illuminated his tracks in the sand, the rest of my team had managed to follow the crocodile into the channel carved in the papyrus. But I had no light. I was several yards behind them, and the sediment they'd stirred up brought my visibility down to zero.

As I stared into the dark passage, every primal instinct screamed at me not to go. *Danger! Turn around! Get out of the water!* The narrow tunnel, lined with tangled weeds, looked virtually impenetrable. Three days earlier, while searching for crocs, Damon, Didier, and Roger had gotten hopelessly lost in one of these maze-like passageways. They had to wait for the river current to clear the sediment before they could find their way back out.

But despite my fear, I had to follow. This was our chance, the reason we were here. So I started to swim forward, following the divers into the tunnel toward the crocodile's underwater den.

The tunnel was about five feet wide—the width of a hippo—and about seventy-five feet long. It was utterly dark; a croc could have been right next to me and I wouldn't have known. Some part of my mind expected to be seized by giant teeth, and for a second I imagined what it would feel like to be caught in those jaws and held underwater.

I pushed these thoughts out of my mind and swam deeper into the tunnel, wondering what awaited me.

A Faint Heartbeat

For much of my life I searched for wildness outside of myself.

As a documentary filmmaker, I made it my mission to seek out the greatest naturalists alive. I met skilled trackers who could read animal behavior in ways that seemed fantastical. I learned about community-based healing practices that gave people a rich, multidimensional view of living and dying. And I was introduced to ancient wisdom about reciprocity that seemed crucial for navigating the future of our species.

During that time I also felt a deep sadness, a yearning. I could not put my finger on the source, but it sometimes seemed as if the more illuminating the subjects of my films, the more my heart ached.

The yearning was more acute around people who knew nature intimately, especially the San trackers of the Kalahari, whom I met when filming a documentary called *The Great Dance*. I was behind the camera, always a watcher, an outsider, while they were in intimate communication with wildness.

I needed to find my own way there, but I was lost—and I sensed I was not alone. All around me I saw people suffering from their disconnection from nature. I sensed intuitively that living in harmony with the wild is humanity's natural state, the state where we are at peace, where we feel most present and alive, and yet so much of the modern world seems designed to cut our species off from the nourishment that nature has provided for all of our existence.

It felt as if there was something inside of me trying to

escape—a wild animal within who could not find its way out. I could feel its faint heartbeat, but I didn't know how to track it or how to free it from its cage.

Awareness came when I was filming Xhloase Xhhokne, a master bowhunter who lived in the arid, thinly populated Central Kalahari of Botswana. I was following him, trying to move with just a tiny bit of his grace. I looked down at my soft hands holding my camera, and then I looked at his hands. His palms were covered with half a centimeter of hard callus, from making bows, from working rawhide, from sustained work in nature.

I had been following Xhloase all day as he searched for food for his family. Midsummer was a difficult time for hunting—temperatures were well over 100 degrees Fahrenheit even at dusk, and there wasn't much big game in the area. Xhloase eventually picked up porcupine tracks and was able to spear one of the creatures just before it went down into its burrow. After the kill, he ate the liver—the hunter's portion—much-needed energy after walking so far. Then he began removing the quills so he could carry the meat home.

Xhloase showed me the hollow tubes of the tail quills that form the animal's rattle.

"Look at the tail rattle," he said to me through his translator, Xamaha. "This one shakes when he is feeling danger, *trrrrrrrr-rrrrr.*"

As he plucked the quills from the animal, the needle-sharp points were sticking into his flesh, but he didn't feel anything because his callus was so thick. His face beamed as he pulled the quills out, and he laughed, not because anything was funny, but because there was a joy inside him.

I'll never forget his laughter, the smile that appeared like the sun on his face, and his strong calloused hands.

I'd spent too much time indoors editing film for up to sixteen hours a day. My hands were soft, my heart was fragile, my smile was fading, and that wild creature was cowering deep inside, with hardly any light to guide its way out. This tameness felt like a kind of death to me—a dishonoring of those who had come before, a dishonoring of my wild heritage.

A denial of my amphibious soul.

An Endangered Species

The word "amphibious" means to live a double life, part on land, part in water, but the word is also fresh with wildness and vulnerability. Amphibians—like the Cape river frog who once took up residence in the pond behind my home—are the most vulnerable class of all creatures on Earth because their skin is permeable and so any form of pollution or toxicity puts them in jeopardy.

Our planet, and all its inhabitants, faces many threats today. Our amphibious soul is not separate from Earth and is under just as great a threat. We humans are permeable too, in our own way. The sufferings of nature seep into our beings, affecting our health, our souls, our psyches. We are all in danger of losing what's left of our wild hearts.

But what is wildness, really? And how can a human being connect to their own wild nature? What would that look like?

Imagine for a moment that you are a human living just a few thousand years ago. Every single thing you eat or drink is completely pure, nothing processed or tainted by toxins. The blood flowing from your stomach to your brain is pure and clean. You've never heard electronic noise or experienced being indoors. The ever-present sounds and smells of the wild—woodsmoke, rain, birdsong—are all you know. When you get sick, you have access to healing practices using wild herbs and medicinal plants.

You are a lifelong tracker, a hunter. Born and raised in a small family group, you are familiar with thousands of wild animals, plants, and trees, and you know every river, bay, and valley of your birthplace intimately. Your whole being is alive and primed, ready and sparking. All of your senses are locked into the wild, and your consciousness and your cognition are firing at their highest levels.

That's a far cry from the distorted view some people may have of our Stone Age past: grunting and hunting prehistoric beasts.

I wanted desperately to tap into this deep intelligence that lives in communion within nature, not at odds with it. But I wasn't looking to leave the present behind and return to the past—rather, I wanted to understand what wildness could look like in this moment in history. I wanted to explore how an embrace of our wild nature might help humanity solve the greatest challenges we face today.

Driven by this deep yearning, I began to take my first small steps toward wildness. I continued to seek out ever more physically demanding and dangerous experiences, like diving with giant tiger sharks and great whites off the coast of South Africa. I'd often wake in the middle of the night after these dives in a kind of altered state, as if my soul had left my body to go back and swim with the sharks. In a kind of twinned consciousness, I'd lie in bed staring up at the dark ceiling while at the same time I was moving through the water, feeling kelp fronds brush against my body, watching the great fish glide by.

But the longing remained.

I wanted out of the cage.

So when Roger approached me with this completely mad idea to dive with crocodiles, I didn't think twice.

Under the Skin of Wild Nature

After what felt like an eternity in that papyrus channel, but was only two or three minutes, I saw a glow ahead and realized it must be the divers' camera lights. As I approached, the channel flared into a cavernous chamber and the visibility went from murky to crystal clear.

Damon, Roger, and Didier were suspended in the middle of the crystalline water while the crocodile rested on the bottom. I took a spot opposite Roger to film. The lair was dark, with fronds of algae wafting in the current and filigrees of light trying to penetrate the thick mass of tangled roots that formed the ceiling of the underwater cave.

Clad in black wetsuits and goggles, bristling with camera gear, we looked like visitors from an alien world as we hovered around the prehistoric beast. The hiss of our regulators and the bubbles streaming around our faces made us look even more alien, likely making the crocodile wary of us. Perhaps that accounted for his lack of aggression; he seemed almost to submit to our scrutiny, allowing us to circle slowly, the camera taking in his imposing stone face, the prehistoric ridges on his back, the intricate spotted scales.

Spellbound in the presence of this enigmatic creature, everything else fell away. I didn't feel the cold. I didn't notice the minutes ticking by on my dive computer. I could see the dragon's horns in great detail, his conical teeth, phosphorescent in the camera lights, clamped in the species' signature grin. Perhaps he was as curious about us as we were about him because, even as we moved closer to film his head, he did not register the slightest bit of agitation. I felt my fear subside a bit, though I did not let myself forget what this predator was capable of.

With our lights illuminating his long body, the huge croc glowed golden. He stayed dead still, watching us, tolerating our presence in his sanctuary. The scene felt hyperreal and hard to fully comprehend. Was I actually there or had I dreamed this entire episode?

He allowed us to remain for a long stretch of time, filming, inscribing wild images against a backdrop of paper reeds on whose pith was written the most ancient texts in history. Eventually, we slowly turned and carefully made our way back out.

"That was one of the great moments of my life," Didier said afterward. It seemed like the wildest thing a person could do— seek out the most dangerous animal on the planet and follow it into its most secret and cloistered place.

Yet even with my heart still racing as we made our way to the surface and climbed back into the safety of the boat, I knew I had not gotten under the skin of wild nature. There was too much distance between the crocodile's mind and my own. I still felt like an observer of wildness—a spectator, a tourist—but I did not, myself, feel wild.

If swimming with the world's most deadly apex predator could not put me in touch with my own wildness, was it even possible?

A Wild Homecoming

Eventually, the wild being within found a way to tell me what I was searching for. It whispered that I would not find my true nature so far from home. It warned me to dispel the old notion that I needed to put myself in great physical danger to know wildness. After all, if danger was enough, then surely swimming with crocodiles and following master hunters on the great tracking hunt would have done it.

I sensed that a different kind of journey was necessary—one that would require me to travel to strange places within, not without. And after twenty-five years on the road, I felt the strong call not to seek out some more extreme location but rather to return to the place where I first encountered wildness: the Great African Seaforest.

A Deep Memory of Abundance

My childhood was spent walking up and down the shore of the Cape of Good Hope and diving into the underwater kingdom of its Seaforest. I glimpsed something in the place where that silver water touched the dry rocks and made them sparkle. It made me sparkle too, perhaps from some deep memory of abundance. It was the last place I felt complete.

One bright afternoon many years after my family had moved away, I returned to the place where my childhood home once stood, invited by the new owner, who also had a love for the ocean.

My old home, the bungalow by the sea with its little porthole for a window that made it feel more like a ship than a house, had been completely demolished by previous owners and replaced with a beautiful wooden home built a little higher on a foundation. A natural stone seawall now absorbed the impact of the giant Atlantic waves that had threatened the house so often in my day. The new owners had changed the home entirely, yet its spirit was still palpable due to their love for this unique spot, this home that lay in the mouth of the great ocean.

While I sipped my tea, standing where my room used to be, I looked out at the same coast I had fallen in love with fifty-three years earlier, with its verdant kelp forest and granite rocks. As a child, I had named each rock—Crack Rock, Big Rock, Crab Rock, Cradle Rock. I looked at each rock and realized they had not changed in any appreciable way in half a century.

As for myself, the changes were undeniable. I was no longer the wild youth who used to comb the beach for treasures. But as I looked out at the place where I was from, the horizon hummed and shimmered, and for a second, I flew across time and felt what it was to be whole again.

CHAPTER ONE

INHERITANCE

I WAS BORN AND RAISED IN THE LAP OF THE SEA.

The tip of Africa, the Cape of Good Hope, is the heartbeat of the world, the coast where humans have had one of the longest relationships with the ocean, perhaps 200,000 years. Africa is the birthplace of us all, and this wild shore is where some of our earliest ancestors first walked. This place is not just my physical home but possibly the ancestral home of every human who has ever lived.

It is also the Cape of Storms, home of giant rogue waves. My earliest memory is of sitting in the bath as a tiny child with my brother, Damon, as a huge wave hit the bathroom door and smashed it open, filling the room with seawater to the level of the bath. I remember the seawater was freezing in contrast to the warm bathwater, and swirling with thousands of white bubbles.

But I made the sea's acquaintance long before the waves came searching for me inside our little home. When my mom was pregnant, she would dive in the cold Atlantic kelp forest without a wetsuit up until the day before I was born. Then, as now, the Seaforest was filled with the magical sound of cracker shrimps. I am still excited each day when I put my head underwater, to hear the clicking of thousands of shrimps snapping their claws, firing bullets made of air.

On the day I came home from the hospital where I was born in Cape Town, my dad placed me in the freezing ocean. Of course, I screamed, but this ritual was part of our family life. Our wooden bungalow was built below the high-water mark, and storm surges used to rush around the house. The house was clad in a kind of waterproof hardboard, but the force of the rain and wind was strong, and my parents had to attach thick wooden boards to the house before storms, to stop the windows from being broken by flying rocks and powerful waves.

I would watch them, Mom in a pair of blue jeans, her long blond hair in a ponytail, Dad shirtless in a pair of tattered rugby shorts, with his preferred brand of cigarette, a Texan Toasted Plain, dangling from his lips.

But the things we build with our human hands are rarely strong enough to withstand the energy of water.

Isopod Intuition

We always knew when the storms were coming, because the isopods told us through their mass migration to higher ground. Thousands of these crustaceans would emerge from the rocky shoreline and crawl into our garden, and sometimes into the house.

"It's going to be a big one," my dad would say. "I hope the sea lice don't block the drain again."

Scientists still cannot determine how isopods know a storm is coming before the barometer begins to drop. In recent years I've come to learn as much as I can about these creatures' inner lives, their mating rituals, and their birthing processes. I have come to love them, though much about their wild intelligence remains a mystery.

I sometimes wonder if we have our own version of that wild wisdom; we've just become too tame to notice it. For 300,000 years of human history, we lived in accord with nature, as free as any other animal. We were nomadic, roaming the land in search of food and water, living in small bands, each member interdependent upon the others. It's only in the past 10,000 years or so that we have become domesticated, spending most of our lives indoors, separated from one another and from the rhythms of nature. It's traumatic: we've lost our ancestral link, our connection to animals, our inborn ability to track—all of those things that keep us healthy in body and mind and spirit.

Yet that wild intuition is still deeply buried inside of us, trying to get our attention, telling us to move to higher ground when the storms are coming.

The Flood

My family always thought that the sea might take our house, but the stream running under the road did more damage than a wave ever would.

One night, I awoke to find my father and mother at my side.

It was still dark and I could hear a harsh wind howling, and rain thrashing against the windows.

"Come, Craig," Mom said. "We have to go."

I looked up at them, still half asleep, unsure of what was happening. Night was a frightening time for me. I didn't like the darkness, and would often sneak out of my bed and into my parents' room when the fear took hold. I was a sensitive child with a wild imagination, and many nights I had the feeling I was not alone in the room. I could feel, and sometimes even see, the presence of shadowy beings moving through the darkness, and when my parents would send me back to my room, I would pull the covers over my head until sleep came. It's not that my parents were cruel; they just didn't understand my fear.

My father seemed invincible to me, and his calming presence that stormy night told me I didn't need to be afraid. Still, I knew something was wrong, and when I looked down I saw what the problem was. The water had gotten in again, and the floor had been transformed into a rushing watercourse.

With my parents on either side of me, my father holding my hand, I stepped into the cold river as my father scooped Damon up with his other arm. The water was already deep, running through the house, cascading down the staircase like a waterfall. As we made our way up the stairs that led out the front door, the water spilled over my feet and I shivered from the cold.

When we got to the road, we saw our old Triumph floating in the current.

I was too young to fully make sense of what was happening, but through the sheets of rain we could see the source of the flood: an empty forty-gallon oil drum had drifted down from the road and plugged the drainage culvert as tightly as a cork in a bottle. And more water had backed up behind the debris-clogged fence that ran along the shoulder of the road.

My dad didn't hesitate. He rushed out into the flooded street and started working on the fence with a pair of heavy-duty wire cutters. It was dangerous work, racing against the rising water to give it an outlet. As soon as he cut through the fence, much of the floodwater drained away. If he hadn't, I think it would have carried our house off.

By the time the storm subsided, the house remained, but it was completely filled with silt, waist-deep up the walls. It took four months to clean the house out and replace all the floors and ceilings. The flood had been so powerful that huge curbstones, heavy stones even my father could hardly pick up, were washed a hundred yards out to sea. When we were diving a month later, we spotted them, looking out of place on the seafloor.

My mom has an amazing memory from the flood of my blue rocking horse being taken out to sea. It must have been visible because of the crude floodlight on the house that used to illuminate the water. I imagine her pausing briefly to take in the surreal vision, a riderless blue wooden horse bobbing in the floodwater under stormy skies.

Treasures

From the age of three, I learned to swim and dive—a burgeoning tracker obsessed with animal life in the rock pools. Every day I'd visit the pools at low tide, excited by the lobsters, crabs, and fish I saw.

In those days, kids were left to roam much of the time, and I loved this freedom in nature, sometimes alone, and often with Damon, who was three years younger. Our grandparents listened with great attention to stories of our adventures in the rock pools and along the shore. This deep care my gran Marjorie and my great-grandmother Gaggie had for us and our childish tales was a powerful catalyst in forging my love story with the ocean that began so young.

"Now tell me everything, Craig," Gaggie would say. "From when you saw the huge crab to when you couldn't get back to shore." In the warmth of her attention, my stories caught fire and flowed with meaning. Fire and water flow differently, but both flicker, and it was that play of dancing light that got me going.

My gran was something of a wild woman herself, a great explorer. She used to go on safaris and come back with incredible stories of being charged by elephants and hippos, and she always brought back a handful of treasures for me to hide away in one of my secret spots. She used to collect semiprecious stones she found in the bush. I still have two that she gave me: an electric-green malachite and a chunk of golden-brown tiger's-eye. The stones feel warm to the touch and always remind me of the glow of Gran's attention.

I remember sitting in our home and staring at the sea for hours, enchanted by its moods and mysteries. On the shore, I found many treasures: wooden carved heads, seal teeth, old steel hooks with Chinese characters carved on the wooden handles. Once, a glass bottle full of letters and foreign currency washed up. A message in the bottle asked us to mail the letters, which we did. The money was for the stamps. It was a miracle the bottle hadn't shattered on the rocky shore.

Calling for Help

When Dad would dive into the water, I'd follow him, amazed at how long he could hold his breath, how deep he could venture, how long he could endure the cold. Neoprene wetsuits had been invented in the early 1950s but it would take some time for them to gain worldwide popularity, so for years my dad wore two rugby jerseys in the water instead.

He was gifted with immense physical strength and was a talented athlete, and he seemed to me a superhero, like nothing could touch him. He was a printer by trade, working from 6 a.m. to 6 p.m. and often weekends, but his passions were the ocean and hiking big mountains.

My mom was a graphic artist and a homemaker. A kind-hearted soul, she was always thinking about how to help others, while often neglecting herself. She came from a family of ocean and beach people—more swimmers than divers. Both my parents knew the sea intimately. While they had little understanding of marine biology, they knew the local species well from encountering them during thousands of dives and swims.

The sea was often treacherous, and my father saved many people who were dragged out by rip currents. Our coast is full of crosses marking the deaths of countless fishermen washed off the rocks by the sea. Dad would swim out calmly and bring people back to shore.

I can remember very clearly the day he saved a man who had been picnicking in the little cove near our house. He had fallen into the sea while walking on the slippery rocks and couldn't swim. In seconds, my father had broken into a strong stroke. I watched his head disappear as he dove under to rescue the man, who had already slipped beneath the waves. Dad pulled him up, brought him back to shore, held him upside down, and pumped his chest, sending all the water rushing out of his body. I watched as the man coughed and came back to life.

On many occasions it was me my father needed to save. I couldn't get enough of the water and would often stay out on the rocks long after the tide came in. It wasn't safe when the sea was rough, and I had to be rescued a few times. I remember sitting on the rocks, fierce currents moving like liquid serpents around me, and calling for help. I remember Dad moving fast and steady through the crashing waves, scooping me up in one arm and bringing me safely back to shore.

Sacrifice

My young life was a series of wild adventures, and so it was a terrible shock when the day came to attend school. I was painfully shy and for a full year hardly spoke a word. I just wanted to be back in the enchanted tidal and kelp forest kingdom. I couldn't wait to get home each day so I could dive into the water and explore the rocky shore. Every day was different, there was always something interesting washing up, and I never knew which animals I'd see. I found school dreadfully boring and predictable in comparison.

Some days when I got home, I'd climb the giant milkwood tree that leaned over our bungalow and eclipsed it in size. I'd tiptoe around the venomous tree snakes called boomslangs that lived in its twisting branches, and bring the artifacts I collected from the ocean and shore—my stones, bones, and shells—and hide them in the tree's nooks and crannies.

Then, when I was ten years old, my parents moved inland to the suburbs so we could be near my new school. Bishops, the school my father had attended, was much better for formal education and sports. It was a private school, and my parents didn't have much money, so the idea was to rent out our bungalow to pay the tuition.

They had so little money that they couldn't afford a moving company, so Dad and his cousin Gregory moved the entire contents of our house by themselves. They worked quickly, running from the house to the car with the furniture and the appliances as if the items weighed not much at all. With Mom helping, they emptied the entire house in a day.

It was quite a sacrifice on the part of my parents to leave this magnificent place so Damon and I could have an education, the best that they could give us. I was sad to leave my sea companions behind, though I don't recall putting up a fight. I don't remember my last swim out to the rocks, or climbing up the giant milkwood tree one last time to collect my treasures.

The sea was such a constant part of my life, I couldn't imagine being without her every day, until it was too late.

Exploring the World

I met new friends at Bishops who shared my love of adventure and nature. One friend, Jeremy, lived near us in the suburbs, but his family had a vacation home a couple of hours up the east coast, at the mouth of the Breede River. We used to take a little boat from his place into the river to dive, or walk along the shore to the ocean.

On one calm day, his father steered their little motorboat to the place where the river met the sea. I remember taking a deep breath through my snorkel and kicking down a few strokes, probably to a depth of about twenty feet, when I felt the presence of something large beside me in the water.

I turned and saw, through the window of my mask, a southern giant octopus, easily as tall as a grown man. Its head was bright orange, the size of a rugby ball, and its arms spanned wider than my arms could stretch. I was quite familiar with octopuses by now—there were many by our old house and I'd been swimming with them for years. But common octopuses were smaller, and when they took hold, it was easy enough to shake their arms loose.

Not this one. At fifteen years old, I was already quite tall and a strong swimmer, yet I was no match for this animal. It grabbed my arms and pulled me deeper into the water toward its den.

I didn't have time to be afraid. I knew I couldn't fight it; it was too strong. And so I did the opposite. I managed to relax and let my muscles go slack. After about thirty seconds or so the octopus let me go—perhaps because I wasn't struggling and it realized I posed no threat.

As I surfaced and swam back to the boat, I felt my arms stinging in the salt water. The octopus had dragged me past sharp rocks and gravel, scraping a layer of skin off my forearms in the process. It took about a week to recover.

Still, I couldn't wait to get back in the water.

My friendship with Jeremy also helped inspire my love of storytelling and film. Jeremy's father had an old VHS camera and a one-to-one VHS edit suite—a very basic setup, but it seemed revolutionary to us. Nobody had a video camera back then, and certainly nobody had an edit suite. It was probably one of the first video cameras in the country. Jeremy's dad used it to film our rugby games. He'd edit the footage and we'd watch the playback so we could analyze our moves and become better at rugby. Rugby was almost like a religion at Bishops.

Jeremy and I were also into comedy, and like most kids, we

thought the stupidest things were hilariously funny. We spent weeks making silly James Bond spoof movies and our version of *Candid Camera* pieces, greatly annoying the neighbors in the process, I'm sure.

There was something magical about holding that camera, learning how it worked, seeing the strange, parallel world in the black-and-white viewfinder. I remember feeling something new emerge when I was behind the camera, a kind of unchecked playfulness that helped me break out of my shyness. Filmmaking gave me a new way to observe the world and make sense of it in stories.

When I left school, it was mandatory in South Africa to do two years' service in the military. Jeremy helped me get into the navy film-and-television unit, where we spent two years honing our filmmaking skills. I was deeply grateful not to be in combat on the border, and because I was stationed at the naval base in Simon's Town, on the shores of False Bay, I was able to dive and swim regularly.

The great bay had a spectacular diversity of animal life. The Cape's African penguin colony had recently arrived and was steadily growing, having abandoned their island home in search of a better food source. I saw huge pods of dolphins, great colonies of Cape fur seals, and many species of whales and sharks. This was the site of the famous *Air Jaws*, a series of fourteen television specials about great white sharks. Hundreds of great whites still populated False Bay at that time and it was common to see them.

After the navy, I left to explore the world outside of South Africa. I moved to London and tried to find a job in the film industry but soon ran out of money and was surviving on one meal of plain oats per day and sleeping on a friend's floor. Right before the money ran out, I got a job as a film editor.

Though I liked the work, loved playing with sound and light, and could completely lose myself in the process of discovering the stories hidden in seemingly disparate images, I felt more disconnected from myself than ever. One day about a year and a half into the job, I glanced in the bathroom mirror and could hardly recognize the person looking back at me.

My skin looked grey, and I felt grey too, in this grey world of tall buildings that blocked out the sun and hid the horizon from view. I longed for the heat and warmth of Africa. Here in London, people hid from the rain when it fell, and so did I, though I sometimes would think of my father rushing into that storm, unafraid, and I remembered the way rainy days didn't keep us out of the water.

No, the rain just meant we'd have the sea to ourselves.

A Tropical Paradise

I knew I needed to be living a different kind of life. With my first wife, Sara, I flew to the Caribbean. Using an old tarp and bits and pieces scavenged from the hurricane that had hit a few years earlier, I built a rain trap and a safe, protected camp in a remote part of the British Virgin Islands. We lived mostly from the wild for four months, diving up to five hours every day in this tropical paradise, amongst giant tarpon fish and moray eels. We ate fish, lobsters, coconuts, and wild fruit. I fell in love with the giant land crabs that lived around our tent, and could not eat them.

When it rained for three days solid, Sara and I shared the last tiny patch of dry land with thousands of insects. We slept wet and bitten for a few days, but eventually the sun returned.

As the greyness of London washed out of me, I felt a yearning to return to South Africa. The pull was so powerful, I was compelled to leave the tropical reefs and forests and return.

It seemed the closer I got to nature, the more I felt drawn back to my deep ancestors, the ones who started us all.

A Real Job

When I finally returned to Cape Town, I was determined to make a living as an independent filmmaker, which many people thought foolish. People constantly asked how I supported myself and what my "real job" was, because filmmaking in those days was seen as a near-impossible way to sustain oneself, especially in South Africa. I had a sense of the films I wanted to make—about African culture and the continent's rich biodiversity—but in the beginning, I took on any film job I could find, from corporate work to low-level adverts, and managed to scrape by.

After about one and a half years of doing this, a strange notion came over me. I vowed only to do work I felt passionate about, with themes focused on the connection between humans and nature.

This seemed like career suicide because there was so little

work in this narrow field of interest. But something in me remained resolute. Somehow I knew that if I stayed true to my vow, I would survive.

This thinking would prove critical in my career and in my life.

AROUND THE SAME TIME, DAMON AND HIS WIFE, LAUREN, RE-turned from their own adventure living on a remote island in Fiji. They had bonded deeply with the local people and had been accepted almost as family, and they brought back fabulous stories of island survival and friendship.

I teamed up with Damon and Lauren, and we began a long and adventurous film career working in twenty-four African countries. For the next two decades, the greatest trackers and naturalists alive let us into their worlds, and we were spellbound.

A trip to West Africa, Mali, was a powerful and sobering experience for our small crew, led by producer Carina Frankal. We arrived at the hottest time of year in the hottest country in the world in an attempt to film special ceremonies performed by the Dogon people at this time to appease their ancestors. I can still taste the arid air of the Bandiagara plateau, where fierce winds carried sharp particles of sand that had us coughing blood for days. Temperatures were over 120 degrees in the day and, even at night, rarely dipped below 100. We climbed mountains in this river of heat to meet the Dogon people, an ancient ethnic group whose culture had endured despite slave raids, religious persecution, and wars.

Entering the Dogon village was like stepping back in time. The entire village was made of dried mud and stone, with houses built on top of one another and nestled in the protection of a tall sandstone cliff. Their clothing was handwoven, their tools made of stone and hand-forged iron. Everything around us had been made by someone's hands. Some villagers had old flintlock muskets and hand-carved wooden spearheads. I was most taken with the blacksmith, who crafted practical and magical tools using a stone anvil and hand-driven bellows. Against the steady drumbeat of the bellows and the clanging of the anvil, he hammered out a cloud hook for catching clouds to make it rain. The jackal diviner held us spellbound with his predictions of future events he'd gleaned from interpreting the tracks left by jackal paws in the night.

Damon and I met the much-feared hunter-sorcerers, three brothers conducting a ceremony—rarely performed today except for tourists—to lead the souls of the dead to their final resting place in the land of the ancestors.

We followed our guides through the maze-like village, stepping from roof to roof. We had to be careful where we stepped; the thatched roofs were old and built to hold people who were perhaps half our weight. High on the sandstone cliff, we came to a sheer rock face that seemed impossible to scale. With the help of handmade ropes, which increased my fear that I would slip and fall hundreds of feet, we climbed higher and higher. There we found a large low-ceilinged cave that contained hundreds of skeletons, grinning skulls, and bone fragments. The more recent dead still wore clothing, much of it rotted away, and their necks and arms were adorned with necklaces and bracelets. I was relieved when we came back to ground level alive.

The next day, we sat in a stone-and-mud arena in the center of the village while dancers performing the ancestral rites wore towering masks made of wood and hibiscus fiber, held in place by their teeth. The heat put me in a kind of altered state. I felt the presence of something profound—a raw energy—and I wondered, when it was my time to meet my own wild ancestors, how I would find them.

Complete Trust

For twenty-five years I lived this grand filmmaking adventure. Damon and I each had a film camera, and I would do the close-ups while Damon shot wide and medium shots. We hardly had to say much because our minds had become fused. Sometimes when I'd look over at him I knew he was imagining an edit exactly as I was.

Our work styles were similar too. We'd both get up long before dawn and work late into the night. The work was very motivating; we were learning so much from the Indigenous people who so kindly let us into their lives. The Dogon and the San were grateful to us for recording their rapidly changing cultures. That was a huge motivation to continue this work, which took many years and delivered not much monetary gain but gave us a priceless life experience.

Lauren used to call Damon and me "Wheresit One" and "Wheresit Two," because we were always losing things in our quest to get the perfect shots. Only Lauren knew where everything was, and she kept us organized, handling all the film finances as well. We would have been lost without Lauren. She was tiny, and it was quite a comical sight to see her commanding two lumbering men, each well over six feet tall. Sara also played a critical role in our film work, supporting me in multiple ways that allowed me to put most of my energy into the creative side.

The powerful thing about working so closely with my brother was that there was complete trust between us, and a strong sense of wanting to share—a rare and wonderful blessing that was deeply instilled in us by our family. We had each other's backs no matter how extreme the situation. And quite often we'd literally be filming back-to-back on foot while a pack of hyenas tried to come in and find our weak spot, or a black mamba rose to strike at our cameras. Once, a crocodile bit the front of Damon's camera.

By some miracle, none of us got seriously injured during all the adventures. But even though it felt as if I was leading a sometimes outright dangerous life, I still had this deep feeling of being outside of myself, lost to my ancestral hearth. And though I spent much time in nature, I did not yet see the wild world as something that could restore me, nurture me, or make me whole.

The Wrong Balance

I was wired, like my dad, to be a workaholic. I remember him putting in long hours, leaving the house early and coming home late. He never took a sick day. That was my model growing up; I had internalized it, and I would never realize the toll it was taking on me until it was too late.

During one of my busiest periods as a filmmaker, I went through the deep sadness of divorce. My son, Tom, was only two years old at the time, and I worried for him, even though Sara and I had a peaceful parting and remained very good friends. After twelve good years together, we both knew our time was up, though as Tom's mother, she would remain an integral part of my greater family.

Filmmaking had taken over my life. Like many artists, I was obsessed with the creative process, and though I knew my family should come first, my work had become the priority. Whatever film I was working on intruded into my dreams and took over my every waking moment; it was all I thought about. I knew it wasn't healthy or sustainable.

I remember one moment, not long after making *Into the Dragon's Lair* and two other crocodile films, when I realized something needed to change.

I had driven from my home in Claremont, a suburb of Cape Town under Table Mountain, to my brother's house in Hout Bay, where we had a studio and an office in the house's lower level. In the office, we had a homemade desk made of two rough-cut slabs from a stone pine, five inches thick and more than ten feet long. We had set these two great slabs of wood side by side, and because of the natural curve of the tree trunk, there was a gap where the boards came together. We had filled the gap with a river of flat stones from the ocean to make a level work surface and arranged our monitors and keyboards and film-editing equipment across its surface.

The desk was a way of bringing the wildness closer, but when my eyes would go bleary from the work and I looked down, I didn't see what it had been, didn't remember its origins as living tree and living rock.

I saw a tool to help me get the job done.

I remember pushing away from the desk after hours of intensive editing. I needed a break, so I went into the kitchen to make a cup of coffee. My brain barely functioned. It felt good to focus on the basic tasks of heating water in the kettle, pouring the fragrant beans into the hopper, and turning the grinder's wooden handle.

Three films at once was too much, but in those days we didn't say no to work. Independent filmmaking wasn't like it is now. The kinds of films we were doing, about natural history, Indigenous people, people living close to nature, were not as popular then as they are now. It was hard to get commissions, so we had to take everything that came our way and operate as if lean times were coming.

Watching the warm liquid begin to drip from the filter into the carafe, I let my mind go blank and just appreciated this moment of smelling and tasting the coffee, of looking out over the little courtyard off the kitchen, of not having to process thirty things at once.

In this moment, this brief respite from the overloaded atmosphere of the studio and the editing desk, I realized I'd gotten the balance wrong. The rush of modern life, the continual need to do more and go further and move on to the next thing, had taken me, as it has taken so many of us, further and further away from myself. I knew if I continued in this way, I would be cheating myself out of something vital to being human: the wildness that is my inheritance, and yours.

What Have We Got to Lose?

Of course, there were also many moments of great joy in those years of adventuring, usually by the sea.

There was the time Sara and I first placed our baby son, Tom, into the sea at Cape Point, a few days after he was born. We held him gently in the water, with the green algae all around. It was a warm day, and though he gasped at the coolness of the water, he didn't cry. As we held him there, part of his belly button, his umbilical cord stump, chose that moment to break off, as happens naturally after babies are born, and it floated away. Watching that tiny part of our son's body being carried out to sea was a powerful experience.

There was the day seven years later when I proposed to my second wife, Swati, by painstakingly creating an underwater message that read "Will you marry me?" out of pieces of broken coral. The fish kept breaking up the pieces and distorting the message, but finally I managed to get Swati out to see it. She was still learning to dive and couldn't take her snorkel out of her mouth to respond. Back on shore, I was relieved to finally get a yes!

And there was the life-changing moment when Swati and I

were taking one of our regular trips to the tip of Africa, to Table Mountain National Park. It was after a particularly grueling stretch of filmmaking, and as we were driving past one of the beaches on the coast, Swati asked me, "What would your dream place be to live?"

I paused for a moment and looked out at the coast.

"My gran used to bring us here when we were very young and we used to have picnics on this beach. This is probably the place in all the world that I'd like to live."

"Well, why don't we get something here?" she said.

"I'm sure these places are far too expensive."

"Let's just drive up and see," she said. "What have we got to lose?"

So we drove into an area of around two hundred houses overlooking the ocean. The mountain behind was so massive it dwarfed the houses. It grew out of the sea in a gentle curve, becoming steeper as it rose. We saw colonies of furry little hyraxes, or dassies, scrambling over the rocks—they looked like groundhogs but were actually distant relatives of the elephant and the sea cow. Far below us we could see the ocean and the kelp forest and huge sparkling granite rocks, with flocks of cormorants skimming in formation over the water.

Eventually, we saw a small plot of land, and when we cold-called the agent, we were shocked at how low the price was. This area was just far enough away from the schools, and people didn't know about it. It was tucked away and not that popular, and the price was lower than houses in the suburbs.

We phoned my dad.

"Try to get a little house that you can do up instead of a plot," he advised. "It's more affordable than to build from scratch."

That night, Swati and I wrote down on a little piece of paper all of the things we would love in a house.

A view of the sea.

A fireplace.

A workshop where I could do my art and keep the artifacts I'd gathered from my travels.

And Swati had always wanted a sunroom.

We took our list of dreams down to the beach where my gran used to take us and we said a little prayer to the ocean. It still felt a bit like an impossible fantasy, but as Swati had said, "What have we got to lose?"

Then we got a call back from the agent.

"There's one place," he said. "Come have a look."

THE LITTLE COTTAGE HAD A GINGERBREAD ROOF AND A ROSE GAR-den. It was quaint, and the couple who lived there had clearly put a lot of care into it, but it looked out of place in harsh, wind-swept South Africa. It was owned by an older Irish couple who, it seemed, had tried to re-create an English cottage. The walls were covered in gold wallpaper, and there was white lace hanging everywhere.

But we could see beyond what was there to what could be— while the house didn't let in much light, I realized we could break down some of the walls to create Swati's dream patio. There was a single garage I could envision using as a workshop. The only thing missing was the view of the ocean.

I walked into the backyard, where there was a tall pine; like the rose garden and the cottage itself, it looked out of place—pine trees aren't native to South Africa.

I began to climb its trunk, and as I got a little higher, I saw it: the glint of the sea. We did have a view of the ocean after all. "My God, this is a beautiful view," I called out to Swati. "We just need to take down this tree, which doesn't belong here anyway."

We put in our offer and it was quickly accepted. We signed the papers that week.

BIT BY BIT, WE TRIED TO BRING WILDNESS BACK TO THIS PLACE. First, we took out every single exotic plant—including the roses and the pine tree—and planted the native vegetation called fynbos to attract birds. Next, we made a beautiful natural swimming pond, filled with aquatic bog plants, rushes, and reeds. Instead of chlorine and chemicals, plant roots and rocks served as filters. The water cycled through the plants and stones the way it did in nature, and swimming in it was like taking a dip in a mountain pond.

We knocked down many of the walls, making one big space full of tall windows to let in the light. We took out the British fittings but kept the pretty bay window. I broke up all the straight lines and brought in a quantity of driftwood I had collected to create a flowing driftwood ceiling. Slowly I started to bring in my treasures from the ocean—sharks' teeth and vertebrae, argonauts and sea fans and hundreds of other shells.

I began the work of converting the cellar into a room for the tens of thousands of artifacts I'd collected during my travels all over Africa, the spears given to me by my San teachers, the stone tools that date back a million years, and the musical instruments from Indigenous cultures of Africa.

Soon the entire house became a kind of living embodiment of nature.

Remember

As I held in my hand the tiger's-eye given to me by Gran some four decades earlier, I realized the house was also a reminder of my wild childhood. All I needed to do was pick up a rock or a shell and I was transported back to the Seaforest in the days before I was pulled away by the demands of life above sea level and into a world that was packaged, mediated, and controlled.

I believe every human has experienced some version of that loss, whether they grew up close to the sea or first connected to a blade of grass peeking through the sidewalk cracks. Though our souls crave communion with wildness, we are a species that has overwhelmingly embraced tameness and "comforts" that anesthetize rather than truly nurture.

But we are made of water, and we came from the water, and when our souls need tending, it is often to the water we return. This shining liquid that makes us feel weightless has a profound effect. If we could bottle this feeling, it would be worth trillions.

As soon as the house was ready for us, I began to embrace a new daily ritual: each morning I would wake early and go for a swim. It was only a five-minute walk to the ocean and the kelp forest, a short bike ride to my favorite swimming spot. From the cul-de-sac behind our house, I could take a path through the forest by a little seasonal stream that flowed in the winter and attracted birds. Along the way, I could pick right off the shrubs the sweet red fruit called num-nums. As I biked, my mind would swirl with memories of my childhood by the sea, and the same curiosity and wonder would overtake me. *What messages will I find in the sand and rocks today? What stories will I bring back to Swati and Tom?*

This homecoming helped me understand that wildness was not something unfamiliar. I knew it each time I had plunged into the water as a child, each time I had climbed a little farther out on the branches of the giant milkwood tree. I knew it even in the fear I had felt when I called out to the strange presence in the dark.

"Who are you?"

Wildness is something we all have touched; we only need to remember.

What wild stories did you tell your elders before you learned the proper way to speak and write?

At what moment did you trade in your own wild inheritance for the promise of safety and comfort?

And was the choice even yours to make?

CHAPTER TWO

COLD

I MOVE QUIETLY IN THE DARK SO AS NOT TO WAKE UP SWATI, ROLL-
ing my left shoulder a few times to soften the stiffness of an old
rugby injury. As I slip out of the bedroom, before closing the door
behind me, I look back at Swati, warm and cozy in bed. She's
an early riser too, though not yet a swimmer, and so these first
mornings at our cottage by the sea, I am on my own.

My morning dive soon becomes a ritual, a meditation:
Place the teakettle under the cool running water, then onto the
stove. While it warms up, spoon out the loose tea Swati brings
back from India. Gather my gear—big thick jacket and pants.
Insulation is crucial when warming up after immersion in cold
water, especially when you don't wear a wetsuit.

Forgoing a wetsuit's warmth and protection was a decision I made early on after my return to the Seaforest, both to
honor my parents, who for many years didn't wear wetsuits,

and because I wanted to break down the barrier between me and wildness.

On these cold winter mornings, I peer out our large sliding window, which looks straight across False Bay, and watch the sun rise over the mountains. Massive clouds fill the sky over steel-grey water. If it's raining, I look for the "rain animals" I learned about from my San teachers, who see the rain as a creature. I look for the rain's hair, long braids of rain that hang underneath the clouds, often curled at the end as the wind sweeps them along. I watch as the rain's body—big, dark voluminous clouds—dances across the churning waters of the bay.

"Don't think of the cold," I tell myself. "Think of the gifts it brings you." Before closing the door, I glance at the stone-clad fireplace, knowing a fire will be waiting for me when I return. That promise of warmth is part of the ritual. Immersion in the cold wild is a threshold to be crossed, a doorway into another world, but it is not punishment.

A Kind of Eden

As I step outside and hop on my mountain bike for the three-mile ride down to my favorite spot, I am hit with the wind, and often the rain. The weather here at the southern tip of Africa is, for the most part, quite pleasant—a Mediterranean climate with mild winters and warm, dry summers. The shrubby vegetation called fynbos—derived from a Dutch word meaning "fine bush"—that carpets the mountains and coastal plains of the Cape is comprised of thousands of different plant species, many of them en-demic, or found nowhere else on Earth.

There are edible plants and thousands of species of shellfish, reptiles, birds, and land animals. You can drink the water straight out of many of the streams without risk of intestinal parasites. There's no malaria. Nature is not always nourishing to humans, but it is here. It's a kind of Eden, a perfect habitat for human life to flourish—and it has. The archaeological record shows that humans have lived on Africa's southern coast for at least 180,000 years, enjoying an unbroken relationship with the ocean. Our early human ancestors were foraging for seafood, giving shells as gifts, and using red ochre some 164,000 years ago.

While the temperature rarely drops below 50 degrees Fahrenheit, the wind can be extreme—in the winter months especially. Many mornings the windchill factor drops to the 30s, and the rain comes at me sideways. After locking my bike to my special rock that has a hole in it, I pull my jacket tight and squint through the piercing rain, and begin the short walk to the crashing waves. The wind is strong down by the beach, where it whisks up sand and tiny pebbles, stinging and scraping the skin. I've learned to look for wind shadows—places where the wind is dampened or blocked by rocks and vegetation. I shelter in these little sanctuaries before entering the water. But there's only so much you can do when the cold comes at you from every direction.

After a bit of stretching on the beach, I speak to the ocean much the way I'd speak to a person.

"Teach me about you," I say softly. "I want to learn."

I know it sounds a bit esoteric, but putting myself in that mindset sets me up to be more open to whatever the ocean has to show me.

Then I put on my mask, snorkel, and weight belt, which allows me to control my buoyancy and float effortlessly on the ocean floor before ascending.

Despite my great respect for this tremendous biological intelligence, my first steps back into the Seaforest after my return to the region have been slow and tentative. I sometimes wonder why I am subjecting myself to such harshness and discomfort.

One cold morning, I stare out at the sea before stripping off my protective layers and pinning them down with large stones so they won't be whipped into the air. The sky is still dark and the water so black it looks like an impenetrable steel shell. In this moment, it feels like one too—closed off and forbidding.

As I peel off my socks, exposing my feet to air that feels close to freezing, I ask myself, "What the hell am I doing?"

Time to Heal

Before my return to the Cape of Good Hope, my life had been cluttered, filled with endless tasks, my to-do list bulging and spilling black sludge. I needed time: time to be without doing, time to allow my conscious mind to be quiet and for the deep innovations trapped in my subconscious to bubble to the surface.

And I needed time to heal.

Each time in my life when I've been most exhausted and worn down, I've felt a deep pull to return to my coast.

Years earlier, when I was at my weakest physically, I spent a short stretch of time by the Seaforest and experienced a near-miraculous recovery. My time in Central Africa had been hard on my body. After twenty-five years of filming, my brain, lungs, and liver were full of parasites picked up in the forests of Gabon and the lakes of Malawi and Rwanda. I'd spent weeks in hospitals, almost dead from cerebral malaria, contracted from mosquito bites.

I remember lying in a hospital in Malawi and seeing black crows at the windows. I still don't know if they were real or hallucinations of my fever-ridden brain. The poor man in the bed next to me died of malaria, as hundreds of thousands of people in Africa still do every year. As I watched his body being carried out of the room—no stretcher, no sheet—I improvised a prayer and waited to be next. I was so far from home, and all I could think about was being back in the nurturing climate of the Cape and immersed in the healing power of the Atlantic Ocean.

I left my bed after four weeks.

I almost cried with relief when I walked down to the beach near my childhood home after I was released. The smell of decaying kelp and salt air was deeply comforting. The sand on my bare feet was a delight. The sea glittered like a thousand jewels, only far more precious.

I lay on my back, floating in the freezing water and letting it soothe me. I stayed in for only ten minutes, but those ten minutes were a turning point in my road to recovery.

While I eventually recovered from the parasites and the

malaria, even a decade later my immune system was still weak and I was prone to frequent chest infections. I had work to do to bring my physical body back to health, but I also had the sense that I needed to undergo healing on a much deeper level.

The glimpses I had received of traditional healing practices among the San and Dogon peoples helped me realize I had much to learn from my forebears—and from nature. Though I didn't entirely understand the deep pull I felt for this kind of healing, I was grateful for it. Many people I grew up with had little or no connection to the spiritual practices of their ancestors—little understanding of where they had come from in the deeper evolutionary sense. Like me, they were never taught as children that even our very recent ancestors had powerful relationships with other realities that could be reached through altered states. This kind of knowledge is not prioritized in a world where the pace is fast and everyone seems to be racing toward more.

But as I was learning, this wisdom is available to us, if only we know where to look.

The Terrible Feeling

I had long had a strange relationship with filmmaking, with the way the work stirred within me a massive energy that was almost always followed by a massive crash.

I called that impossible up and down "the terrible feeling."

The terrible feeling is this exhilarating and frightening force that visits many filmmakers and no doubt many others who are passionate about their work or who feel driven by something larger than themselves. This mysterious energy seemed to come from somewhere outside of me—a manic muse who granted my wish for tremendous energy and stamina, allowing me to work with almost no sleep in extreme heat or cold for twenty hours a day.

But the bargain came with a cost: once in control, this capricious muse hoodwinked me into thinking that whatever I was working on was the most important project in the world.

I had to make it happen no matter what.

At first, I thought the feeling was my ego trying to prove itself—*Do I have what it takes to be a filmmaker, to express the deep passion and emotion I feel for nature?*

But after proving that with my earliest films, the terrible feeling was still there.

This torrent of energy felt ecstatic at first, but it slowly and steadily drained the fuel tank. Often I felt fine on the surface and could work in this way for months, even years. But looking back, I see that I was chipping away at my health reserves because I was so obsessively focused, not sleeping enough, pushing forward at all costs.

I usually felt the shift coming over me at the beginning stages of a project. It started as pure exhilaration when a piece of the story I wanted to tell revealed itself to me: *This is it! This is the story that will take people inside nature, help them better understand humanity's origins, maybe even move them to take bold action to cherish and protect wildness.* And then the muse took over, flooding me with the energy to do the work and allowing me to experience remarkable things.

I remember one such episode quite clearly.

The Living Past

I was driving my old battered Toyota Hilux 4×4 toward the Nyae Nyae Pans in Namibia, home to Ju/'hoansi[1] San people.

The truck had made many such treks by then, and it was rigged for rough terrain with a grille on the front to stop the grass seeds from blowing in and clogging the radiator. It had a huge roof rack mounted on top for sleeping in the bush, and at night we'd put the hood up so lions couldn't jump up onto the roof.

But the old truck had seen better days. There were holes in the floor from hitting trees in the bush, and it was constantly breaking down. The only reason it was still running was through the wizardry of a German mechanic named Gunther, whom I'd called upon many times to put the damn thing back together.

Still, I couldn't afford a new truck. And it carried the memories of so many journeys: of blown-out tires in the Kalahari bush, of dropping into three-foot-deep aardvark holes and being cranked out with a Hi-Lift jack. I carried the memories too, needing only to glance at the snake skins covering the dash, the big baboon skull perched against the windshield, its eye sockets reddened with ochre, to be whisked back in time.

How the truck was still running was a kind of miracle.

How I was still running was a miracle too.

THERE IN NAMIBIA I FELT THE FIRST STIRRINGS OF THE EXHILARAT-
ing force as our small crew began to cross into the beautiful flat
landscape, with its ancient baobab trees and vast salt pans—
flat areas where water had evaporated, leaving behind salt and
other minerals that often shine white in the sunlight. One of the
trees had a giant cavity, a hollow big enough for ten people to
fit, where the trunk had been split open and then healed itself.
I remember hiding inside the tree cave, looking out at a herd of
foraging elephants, rumbling and flapping their great ears as they
conversed with one another. The tree acted like a giant resonator,
amplifying the deep sounds the elephants were making.

We were doing a film called *My Hunter's Heart*, and we'd been
invited on what could have been one of the last giraffe hunts with
poison arrows. Our guides into the living past were San trackers
who still knew how to hunt in this ancient way, who still had the
knowledge of a weapon at least 24,000 years old, possibly older.
The arrow's toxin is created from the larvae of the *Diamphidia*
beetle, found in cocoons that are dug up. The poison is then
mixed with saliva and used to coat the shaft of the arrow.

Damon and Lauren were in Antarctica at the time, so I was
doing this shoot with two young people on internships. My
brother and I had always worked together so effortlessly and I
desperately missed having him there. The interns worked hard,
but they suffered from the blazing heat, long hours, and lack of
sleep.

It was extremely difficult to keep up with the San hunters
during the grueling hunt, a several-day foot chase under the

burning sun that culminated in the great giraffe dance, a ritual ceremony honoring the animal's sacrifice.

That evening the full force of what we were witnessing took over. I remember sitting by a little fire, with three of the hunters sleeping around me and the massive giraffe carcass resting on the ground next to us. I could hear hyenas hooting and grunting and a leopard coughing in the darkness beyond the flickering firelight. Human beings and our distant ancestors have been hunter-gatherers from our beginnings, as far back as two million years ago—but I knew I was witnessing the end of that greatest era of human endeavor.

The next day I went to the village and danced through the night and into the following day as the people sang and celebrated the success of the giraffe hunt. These people were so giving, allowing us a last glimpse of the hunting and gathering spirit that has carried us from our deepest times.

I was grateful that I was able to experience something that very few outsiders would ever see. I also knew I had pushed myself to the brink, and that the return home would be difficult.

INEVITABLY THE MUSE TIRED OF ME, AFTER TAKING AS PROFUSELY AS it gave. When a shoot ended, after running on that relentlessly fast engine, it was excruciating to try to return to life as a normal, functioning human being who could be pleasant to others. And I still faced a three- or four-day drive to get home.

Actually, the tedium of driving for twelve hours each of those days was soothing to my brain. The old truck had a cassette player, and the interns and I listened to my tapes of the great Johnny Clegg and Juluka. My favorite song was "Impi," about a celebrated battle during the Anglo-Zulu War—the Zulu army defeated the British, but both sides suffered devastating losses.

After days of driving, we were about halfway home when we came upon a funny little restaurant in the middle of the desert called Windpomp, which means "windmill." Our film crew had spent days eating canned or rehydrated food and sharing whatever we had with the San as they shared whatever they had with us. One morning I'd eaten some of the bone marrow from the huge leg bone of a giraffe. Living wild, the body naturally craves fat, and we appreciated them sharing this precious gift with us. But on the whole, we'd all eaten very little and lost weight. So it felt surreal to sit in this restaurant and order from a three-page menu, to have someone serve us food. I ordered mutton chops and french fries and Greek salad and devoured everything on my plate. Of course, the food was much too rich after our time in the wild and I spent the next day throwing up out the window of the truck.

ONCE I WAS HOME, THEN CAME THE CRASH.

At first, huge waves of relief came over me and I wondered why I ever invited that manic muse in. But with the relief came exhaustion, and an anticlimactic kind of emptiness. I just wanted to sleep for days and rest my body and mind until my strength returned. Walking around in a kind of liminal state between sleep and wakefulness, I'd hear Swati or Tom say something that would momentarily snap me out of my stupor, and only then would I realize days had passed without me truly connecting with them.

BATTERED AND EXHAUSTED BY MY EXTREME RELATIONSHIP WITH the creative process, I vowed to enter the great Atlantic Ocean each day. I made the commitment to dive every day for ten years, no matter how much the wind and cold urged me to stay in bed—I knew it would take at least that long to really immerse myself.

And I craved that immersion: I'd observed and learned much about nature and tracking from the people in our films, but I hadn't actually lived a truly wild life myself. I needed to somehow get inside nature, feel her inside me, know her, instead of merely observing and studying her from the outside.

I hoped this commitment might help quell the terrible feeling. I wanted to transform my manic muse from a dominating tyrant into an inspiring creative force that would still grant me stamina and vision but not break down my health or distance me from my family.

What was this force and why did it drive me so hard? Was there a way to do my work without running myself into the ground? And was there a way to feel that aliveness but in consistent, nourishing doses?

The only way I'd ever known how to get the energy back was with the next film.

I knew I had to break this debilitating pattern. And somehow I knew the ocean would help me do it.

The Cold Cure

From a childhood spent following my parents into the cold ocean with no wetsuit, I'd learned that days when you got cold were good days.

I scoured the internet and read everything I could find to learn more about the science of cold immersion. I read about an ancient Tibetan practice called Tummo, in which monks drape themselves in soaking-wet sheets in the freezing cold to expel negative thought patterns. As the monks perform meditative breathing, their own body heat warms and dries the sheets, making them impervious to the cold. It sounded extreme, but I was fascinated.

Everything began to fall into place when I discovered the work of Andrew Huberman, a Stanford professor of neurobiology who had studied the effects of deliberate cold exposure on building physical and mental health.[2] Listening to Huberman's podcast, I was excited to hear scientific explanations for many of the extreme changes I'd experienced diving in the cold waters of the Great African Seaforest.

I LEARNED THAT ENTERING WATER 60 DEGREES OR COLDER CAN cause a powerful surge in neurochemicals like dopamine, adrenaline, and noradrenaline. So that's why I felt so good in the cold—my brain chemistry was just rocketing! The dopamine made me feel good, motivated me, and connected my mind to reward and pursuit. And the overall natural chemistry effect increased my levels of happiness and well-being. Such is the power of cold stress.

Intuitively, I worked to actively relax my body and mind while entering cold water, to prevent too much cortisol from flooding my system.

Huberman even suggested engaging in cognitive activity, like math problems, while in the cold, in order to teach the prefrontal cortex to stay engaged during high levels of stress. He quoted a study that pointed to maximum dopamine release when cold exposure was preceded by caffeine and intermittent fasting. I found that fasting also improved my breath-hold very noticeably. On days when I'd had caffeine, however, I noticed a decrease in the amount of time I could hold my breath.

My exploration also led me to speak with Wim Hof and his son Enahm.[3] Wim had broken several cold exposure world records and had developed a popular cold therapy technique said to strengthen the immune system, among other benefits.[4]

Of course, one of my greatest teachers was the cold itself. I tried all forms of getting cold, from cryo chambers to draping wet sheets over my body in icy winds to iceboxes, but nothing beat the raw majesty of immersion in the ocean.

I SPENT MANY OF MY FIRST EARLY-MORNING DIVES SHIVERING UN-controllably. The cold stung my hands and feet, and the icy wind and freezing water seemed to suck the heat from my body.

As I've exposed myself to cold daily over many years it's become easier and easier.

The one kind of insulated gear I've embraced over the years is a neoprene hoodie to protect my ears. My doctor urged me to wear one after discovering that my ear canals were largely blocked by a bony growth from prolonged cold exposure. If the canals became any more blocked, he warned, I would need to undergo a painful operation.

But during those first few years of cold water diving, I wore no protection at all.

My studies also taught me that the human body contains several different types of fats. Most is white fat (which is actually yellowish), which stores calories and tends to accumulate around our midsections and make our pants uncomfortably tight. A small amount is brown fat, which regulates our body temperature and keeps us warm, creating heat by helping us burn calories. We also have beige fat, which is made up of white fat cells that have gone through a "browning" process, encouraged by good nutrition, exercise, and exposure to cold temperatures.

We are born with stores of brown fat around our shoulders and backs—it's what keeps newborn babies warm since they're unable to shiver. As we get older, if we don't expose ourselves to cold stress, we lose that brown fat and start to shiver in response to cold. You may remember rarely feeling cold as a child? That's because your internal brown fat heater was keeping you toasty—and you probably jumped up and down a lot too.

Despite the discomfort, the cold water filled my brain with

feel-good chemicals that kept me coming back every day. The extreme effects of well-being, happiness, and motivation would last for many hours after exposure—often the whole day and night.

Slow

Eager for improved health and a deeper connection to the wild, I wanted nothing more than to immerse myself fully in the sea. But every time I pushed it too far and stayed out in the cold too long, my immune system would collapse and I'd get a chest infection. I had to go slowly, inching myself back to health.

One of the powers of cold water is that it does not allow us to rush. Slowly my body began to build the brown fat heaters that would eventually keep me warm in cold water. Slowly my mind began to find its way back to my authentic self.

The process of healing in this way was like swimming through murky water. At first, my inability to see brought on anxiety. Then, as the water became clearer, my mind also became sharper, the fear dropping away.

Before long, my body and mind were working in unison.

Where before there was dullness and a tendency toward negative thoughts, the new feeling of cognitive and physical clarity was electric, pulsing with joy and life.

Whenever doubt crept in, I reminded myself that my deep ancestors got cold and wet every day, especially during the winter, and that my body and mind were, in fact, expecting the same thing. Depriving myself of this innate evolutionary experience was depriving myself of vitality and well-being. Constant daily exposure to cold water and wild animals lets the primal mind know it's alive, stimulating and strengthening the immune system.

I also reminded myself that my father regularly dove into the freezing Atlantic Ocean without a wetsuit, always coming out relatively unscathed. I do have one faint memory of a time when he stayed in too long and couldn't get warm afterward. My mother gave him a glass of brandy (which of course we now know actually causes body temperature to drop) and ran him a bath with water from the kettle so hot that Damon and I couldn't put our hands in it. I watched the steam rise from the tub and my father shivering in the scalding water he couldn't even feel.

It took him several hours to get warm again.

Still, he never worried much about it. Even today, at eighty years old, he can easily dive for an hour in cold waters.

These memories would come flooding into my mind with each dive into the cold, and then freeze in time as they'd be replaced with pure, clear awareness of the world around me. The cold is only painful in those first few moments before the body adjusts. On the other side of that discomfort, wonderful experiences await.

Curious

Often these experiences in the cold seemed to whisk me out of my body altogether. One morning my full attention was captured by a magical being: a Cape clawless otter. Wild otters are some of the shyest creatures on Earth, and it's rare to see, let alone get close to one. This otter approached me from behind—most animals feel safer away from the mouths, hands, and claws of other beings, which they see as weapons. Knowing that, I didn't turn to face the otter head-on but stayed still, floating quietly on my stomach and watching him in my peripheral vision.

As the otter brushed against my feet, an electrical pulse rushed through me. My stillness encouraged his curiosity and he swam closer, approaching my front. Eventually I got a full view of his face: lively, expressive eyes, rounded ears set far back on a sleek head, mouth bristling with long white whiskers. Then the otter did something that shocked me: he came even closer and reached out to stroke my face with his dexterous paw, while looking into my eyes. I was flooded with an overwhelming mix of emotions: love, gratitude, and a bit of confusion. As I felt tears well up, my mind filled with questions.

Was it simple curiosity, or was something else going on here, some deeper bond between this animal and humans? I willed myself to rest in the mystery and not try to explain it.

After some time with this creature, I was so full of joy I needed to get out and lie on a rock. The otter stayed close to shore, beckoning me back into the water with a high-pitched call, perhaps inviting me to join the hunt for shellfish. That might seem far-fetched, but throughout history there have been quite a few examples of humans and animals, including dolphins and orcas[5], hunting together.[6] Sure enough, after I relayed to her my experience with the otter, Swati told me about a handful of humans who still hunt with otters in Bangladesh.[7]

The thought of such rewarding moments kept me going

through weeks of long, hard diving. I was learning that if I dove every day, I would see and learn extraordinary things.

I would also witness how the most ordinary things can become extraordinary.

Connecting

Submersing myself in the cold was not only restoring my health and vitality but also giving me more energy for the things that mattered most. I found myself connecting with my son, Tom, in a way I never had before, often taking him on my swimming adventures. When he was a child, his brown fat heaters were still cooking and he was nearly impervious to the cold.

Like me, Tom had been placed in the ocean as an infant, before he could even walk, or talk. It filled me with joy to see that he felt as at home in the water as I did. In the early days before I understood the power of cold, Tom wore a little wetsuit. When he got a bit older, he wore no protection and was at ease in huge waves crashing around us as we glided past jagged rocks and through underwater tunnels.

Many days, we explored the Seaforest together, but I would also put him on my back and take him on hikes up and down the coast. I wanted my son to have a sense of wildness, an awareness of his ancestral origins. We visited cheetahs, lions, and honey badgers at the Jukani Wildlife Sanctuary. We camped out in the Kalahari. Once I took him into a cave in the Cederberg, a region renowned for its ancient San rock art, and lifted him so he could take a closer look at the five-thousand-year-old cave paintings of half human, half animal forms called therianthropes. Of course, at that age he was more interested in the shiny watch on my wrist.

One morning, while walking along the shore not far from the cottage, Tom and I found an old glass medicine bottle probably half a century old. Tom carefully removed all the sand, and inside we saw two bivalve shells considerably larger than the mouth of the bottle. They must have grown and died inside the bottle, their whole lives spent inside a tiny glass vessel floating in the vast Seaforest.

As we sat on a rock, I began to tell Tom stories from my childhood, of our small wooden bungalow, and the night of the great flood. I told him about the glass bottle that had washed up stuffed with letters and foreign currency, and the note my parents received years later from a couple thanking them for posting their family letters, explaining that they'd tossed the bottle off their passing ship when it was unable to make land.

The bottle Tom and I found that day was both a sanctuary and a trap. It reminded me that the soul, too, can feel safe inside the small world of the ego, seemingly protected from the great existential unknown.

The tame world can distort our priorities, leading us to place too much importance on gratifying our egos. In the wild world, no one individual is more important than any other. When a San hunter brings a big animal home to feed the village, the people tease him: "Why did you even bother? Why'd you bring us this old bag of bones?" They do this because they know the ego is a dangerous thing. They've had thousands of years to see how placing some people above others creates envy, jealousy, greed. And when our tame-world egos and desires grow too big, we begin to fight with one another and we devour our own home, our Mother, our sanctuary.

Like many parents, I worry for my child's future.

A Patient Teacher

To my surprise and delight, by the time the warmer months arrived after that first winter in our new house, I found that I had begun to miss the cold. Hungry for the benefits the cold had already begun to bestow, I bought an old chest freezer and modified it into an icebox by waterproofing it with marine-grade silicone, filling it nearly to the top with water, and putting it on a timer for four to five hours a day. That was long enough for a thick layer of ice to form on the surface, which I would crack into chunks before climbing into the freezing water.

Even after months of cold-water swimming, getting into the icy water was challenging. I had to force my mind not to think, to immerse and then relax my body. I learned to take calm, slow nose breaths using my diaphragm.

The first minute was the hardest, and the two that followed often felt excruciating. But I found if I could breathe through my nose into my diaphragm and stay calm that first minute, my body would settle down, and after three minutes, the intense pain would disappear. Sometimes I'd close the lid and bathe in the ice water in the pitch dark.

After nine minutes, numbness would set in, and I needed to

be careful to ensure I was being safe and not at risk of hypothermia. The trick to doing this safely was to increase the time very slowly each day and not to push it—especially if I had not slept well or had other life stresses.

The cold is a patient teacher—taking your time and watching how the body responds is key. The first few plunges were very difficult for me, as the pain was so intense, but I was amazed at how quickly my body adapted. After about five immersions, the pain dramatically reduced, and after another three, it disappeared completely.

I'd typically stay in the water for about ten minutes and sometimes as long as twenty-two. Afterward, my skin would be tight and bright red, and it would feel as if I were wearing an ice cloak over my back—a now-familiar feeling I call "ice back."

I was slowly realizing the great power of cold: it could almost instantly change my state of mind, it was improving my health, and building up my stamina, allowing me to get closer to nature.

The cold can be dangerous, of course, and people have had good reason to dress themselves in warm hides and keep the fires burning into the night, but it also seemed to be giving me a tremendous sense of freedom, and bringing me closer to my authentic self.

Partners in Cold

For many months, I dived alone and had our little stretch of ocean all to myself. The solitude was precious—it was easier to get closer to the wild kin when I was alone, because alone I was quieter and less threatening, and my full attention was with the animals.

In time, however, I began to discover how rewarding it was to share this practice and joy with human kin. When I told friends, and friends of friends, about my immersion into the cold and my thrilling encounters with the creatures of the Seaforest, many were curious.

When they asked how they could begin to develop the same relationship with wildness, I gave them all the same recommendation: "Go and do ten dives by yourself without a wetsuit and then come back if you're still interested."

Ninety-nine percent of people never went through with it, but the few who did became great friends and collaborators.

One of them was Pippa Ehrlich, probably the most determined person I know.

When we met, Pippa was working as an environmental journalist. Though she was quite young, she struck me as an old soul. Perhaps that's why she has such a funny sense of time—I often tease her about being a "tidsoptimist," a word of Swedish origin that refers to someone who always thinks they have more time than they do.

In fact, Pippa was late to our first meeting, something that usually annoys me, but she quickly won me over with her obvious deep interest in nature. Pippa had been diving in the Seaforest since her early twenties and exploring the kelp forest for almost ten years before we met. She was athletic and a strong swimmer. She wanted to learn to adapt her body to the cold and to track animals underwater, and she'd been thinking for years about early humans and how to get closer to the wild world.

I could see Pippa had a sharp mind, and I sensed an untapped talent. She became a good friend, and a kind of surrogate daughter to Swati and me.

Diving in the wild underwater forest without a tank and regulator—and eventually, without a wetsuit—was, she told me, utterly freeing, especially for a mind so often filled with anxious and disorganized thoughts. It allowed her to get closer to animals in a way she had not thought possible and to better connect with nature. This created a focus in her life that became a powerful pathway to creative discipline.

After we started diving together, I had a strange hunch that this special person could help me with a story taking shape in my mind about the relationships I'd begun cultivating with the animals of the Great African Seaforest—a story that would eventually become the film *My Octopus Teacher*.

The Danger Zone

Curious to test my limit with the cold, my impatience took me deep into the danger zone. The icebox had been leaking and I needed my cold fix. In desperation, I roughly crammed some black plastic sheeting into the freezer to plug the leak.

Lying in the freezer wrapped in black plastic one evening, ice floating around me, I felt a bit like I was encased in a body bag. I expected the sharp ice crystals to pop the plastic so I kept very still and began to observe my mind.

It was busy, full of chattering thoughts rushing around. My elbows were aching badly and my right foot felt like someone was driving a nail through it. I pushed on because I knew relief would come. Sure enough, after about three minutes the ache subsided and I began to relax as the crushing cold enveloped me like an old friend.

As I hit the ten-minute mark, I felt the most wonderful thing. The clamor of thoughts started to quiet, as if my "daily chore" mind was being eaten by my deep primal mind.

Chattering monkey silenced by unmoving, coiled reptile consciousness.

At twelve minutes, all daily thoughts had disappeared into bliss, and rising in their place I felt my Mother Mind. I felt all those who had come before me and the great planet resting in the giant hand of space—deep, calming, open space, with just the sounds of the ice clinking together, the sensation of air entering my lungs as I breathed in and out. It felt as though I'd entered a raw state of being, the under-architecture of the human mind.

A chant began to echo inside my head, a rhythmic string of vowels that had no meaning and yet calmed me. Instinctively, I started to chant aloud. It was not something I had ever heard; it came from the back of my throat and vibrated soothingly in my chest: *ay oh ay ee oh ay, ay oh ay ee oh.*

Somehow, I felt warm and deeply comforted in the icy embrace of the freezing water. Even after close to twenty minutes, I felt perfectly fine, but I didn't know the extreme cold well enough yet, so I erred on the side of caution and wriggled out of my body bag.

My coordination was impaired, and my feet felt like all the bones had come untangled, but the looseness was not unpleasant. I felt the familiar sensation of a cloak of ice pulling on my back and neck.

My mind razor-sharp from the cold, I looked around and saw all the plants and objects in my garden with crystalline clarity, as if I'd just cleaned the lens of a camera. The foliage looked vibrant and jungle green, the water in the pool mercury blue. A few minutes later, I felt the "after drop" sensation, as the blood from my extremities went to my organs to protect them from temperature drop, leaving my hands, feet, and skin icy cold. But I felt clear and elated for hours afterward.

AS USUAL, I WAS FASCINATED BY THE EDGE, TO SEE HOW FAR I COULD go. So the next day I went to thirty-one minutes in the icebox. At first, I felt fine. My feet were aching, but the rest of my body felt so warm I was tempted to do another ten minutes, as I wasn't even shivering.

But I had mistakenly thought that shivering was a good indicator in this extreme cold. After thirty-one minutes, I climbed out to find I could hardly walk. I was so stiff it felt like I was walking on raw bone. My vision went blurry for maybe twenty seconds, then returned. Spending this long in the cold was clearly a big shock for my body, but after I warmed up, I felt the familiar high, and all the anxiety disappeared.

Maybe I could do an hour?

The next day, however, I woke up feeling exhausted and as if I had caught the flu, with a cough and the beginnings of a cold sore on my lip. I had pushed my immune system too far.

It took me three days to recover, and during that time, I learned to have more respect for the extreme cold. I should have slowly increased from twenty minutes, perhaps adding a minute every few days. To jump from twenty-two minutes to thirty-one minutes was really foolish. The body responds best to a slow increase in cold stress.

Coming from a hot, subtropical Indian climate, Swati was at first mystified by my fascination with the extreme cold. She couldn't understand how I could tolerate it, and she was upset when I pushed myself too far inside the freezer. She herself was very sensitive to the cold—even in a wetsuit, her hands and feet would get painfully numb—but when she saw how it was benefiting me, she started getting curious and decided to try her hand at cold training. She persisted, and today she has amazing endurance in the cold water.

The Wild Choice

One rainy winter, an icy northwest wind howled across the bay. I had spent the morning trying to master my new open-water kayak with great difficulty. I couldn't get the balance right, and the kayak kept flipping over and spilling me into the water. I practiced getting back into the craft in deep water, and by the time I returned to shore, I was cold and stiff. My mind began to wrestle with itself: one half wanted to go have a hot bath and relieve my hunger; the other half wanted to go diving in the clear, cold water.

I refused to give in to the tame mind, as by then I knew I would feel better after the dive. I also knew from my cold training that I wasn't in any kind of danger; it was just a decision between immediate comfort and putting the warmth off a bit longer to feed my wildness.

I knew this thread of wildness was keeping me feeling alive.

And so I put off the walk home and decided against putting on the short wetsuit I keep for long kayaking sessions that also acts as a life vest.

About ten minutes into the dive, I got a delicious sense of being warm in the cold embrace. I relaxed, and thoughts of the hot bath and meal were lost in the swirling splendor of the kelp forest.

As I stepped out onto the sand and into the wild wind, I realized all my feelings of cold had left me, and I wondered how much of our sensation of discomfort is actually in our minds. I walked home on the little path through the forest, and enjoyed a hot meal by the fire while the freezing wind continued to howl outside.

I WAS STILL SITTING BY MY FIRE LATER THAT AFTERNOON WHEN I heard a knock on my door. It was Pippa, along with my friend Jannes Landschoff, each with their diving fins sticking out of their backpacks.

Like Pippa, Jannes is about twenty years younger than me. We met when he was working on his PhD in marine biology. Born and raised in Germany, he looks like a surfer with his rumpled sun-bleached hair. He's got a brilliant scientific mind, and he's taught me a lot about marine biology and nature.

Early in our friendship, I was the one encouraging Jannes and Pippa to embrace the cold, but it didn't take long for them to start coaxing me to come out. I'd hear a bang on the gate, and there they'd be, shivering after a dive, desperate to get into the heat of the sauna. Without a word, they'd dash by me and into the sauna, like two eager kids with no time to spare for old Dad.

That particular day, they wanted to go diving and invited me along. I mulled it over. Often I will not keep pushing the cold barrier all day, but as I looked through my window down toward the Seaforest, I could see a patch of clear water. It called so strongly to my heart that I decided to go.

We got to the water's edge and sat in a small nook protected from the fierce wind. Noticing the massive grey clouds hanging over the ocean, I moaned a little about the cold.

"Put on your big boy panties and take your spoon of cement to harden up," Pippa said, laughing. I'd often given her the cement line in the early days, but now the tables had turned.

"You deserve it for what you've put us through," Jannes teased, his face breaking into a grin.

Somehow as we faced the cold and mighty ocean, there was more room for laughter than I ever imagined possible. As I followed my friends down to the water, I reflected on my first solitary plunges into the water. Now things had come full circle, with friends leading me bravely and boldly into the cold.

My arms were so tired from the morning's kayaking venture and swimming that I put on fins and a weight belt. Our plan was to move along the shore looking for juvenile giant limpets—mollusks that have formed a deep relationship with the algae in the area.

But shortly after entering the water, I spotted an unusual sight: a helmet snail climbing a rock. I normally only see these animals at night because by day they tend to hide under the sand. Jannes continued along the coast in search of the limpets, while Pippa and I hung back to watch this oval-shaped creature.

The snail glided on its foot up the rock face. And then its destination became clear: it was launching an attack on a spiny Cape sea urchin almost twice its size. As soon as the urchin noticed the approaching predator, it began to creep away over the rock, but the snail gave chase.

It might be hard to imagine a soft-looking snail threatening an urchin armored with needle-like spikes, but this was no ordinary snail. The helmet snail is a fearsome predator who possesses terrifying chemical and physical weaponry.

The helmet reared up onto the back of the urchin and latched on with its sticky foot. It then began to use its weight and twisting motion to wrestle the urchin off the rock. Both animals fell to the seafloor, and there the helmet released a transparent sticky gel, a paralytic enzyme that incapacitated the urchin's spines. This weapon is an invisible shroud, detectible to us only because the web of goo captures bits of sand and shell. The snail then slid underneath the urchin and painstakingly managed to flip it over, leaving the urchin's vulnerable mouth open to attack.

Pulling apart the now-useless spines, the snail dropped acid on the vulnerable spot between the urchin's hard shell and its teeth. While the acid dissolved and softened the shell, the helmet used its toothed tongue to drill a hole large enough for the proboscis to reach into the shell and scoop out the nutritious insides.

I felt sorry for the urchin even though their numbers are plentiful—overfishing in the Seaforest has wiped out many of their major predators. But I also marveled at the genius techniques the helmet snail used to overcome its prey.

The Costs of Comfort

Only when this battle was over did it register that I'd been keeping still for some time watching the drama play out, and in the meantime the freezing wind had drawn much heat out of my back.

I was cold.

The three of us returned to the shore, where we were lashed with a biting wind. As we wrestled our clothes on with numbed hands, we shared the adventures of the dive and then began the trek home to our cozy homemade sauna.

As the sizzling hot rocks brought heat back to my body and memories of the day warmed my mind, it struck me how much stronger I'd become since moving to the Seaforest. No longer was I plagued with frequent chest infections. I was a more confident swimmer. The cold had nursed me to superhealth and brought with it a new sense of clarity and peace.

As with any intoxicating substance, I've learned I must treat the cold with respect. My immersion in the cold has led me to take unnecessary risks and endangered my life. If I'm honest, I'm a little frightened by the total lack of fear created by the cocktail of cold-enhanced brain chemistry.

But the benefits are undeniable. My journey to wildness has been a literal dive into the cold, but it has also inspired an examination of the costs of comfort. I've begun to ask questions like: What happens when society embraces a climate-controlled existence over an invigorating icy plunge? Can we learn to sit with our discomfort a bit longer? What insights await us on the other side?

Our aversion to discomfort causes us to miss out on so much. After all, the pain is fleeting, the threshold is in many ways illusory, and what lies on the other side of those first three minutes is a lifetime of wonder and discovery.

For hundreds of thousands of years, our human ancestors endured extreme temperatures with little protection. Our ancestors survived by sleeping in groups, sharing body heat. Only together could they keep the cold at bay, and only together could they forage and hunt to find enough food to survive. No one could survive alone for long. Tapping back into our ancestral wisdom is not a solo task but a shared journey that we can do together. The creature comforts we have today that have become so unsustainable they endanger our planet also serve to separate us from one another. Sitting alone in our overheated homes, buying whatever we want with a click, we lose ourselves in our screens while the wolf scratches at the door, beckoning his human kin to come back outside and join the wild ones.

The warmth of the fire was soothing and pleasant, but as I began to feel sleepy, I realized it had burned away that feeling of wild aliveness that the cold gives so freely.

I knew that to stay here too long was to risk everything.

CHAPTER THREE

TRACK

PICTURE A FLAT, SANDY SEABED IN A STRETCH OF THE SEAFOREST where the water is about fifteen feet deep and a huge rock rises all the way to the ocean's surface. Where the seabed hits the rock there's a small cave about the length of your body that tapers to a narrow fissure not much bigger than your arm, making it the perfect home for cuttlefish. Added protection is provided by the kelp, and the sand here is nice and clean, reflecting light into the cave.

I call this Cuttlefish Cave, and it's one of my favorite places. One morning I brought my friend Aaron Friedland, a Vancouver-based explorer, lecturer, and podcaster, to see it.

Sure enough, five patchwork cuttlefish were floating gently in their little hideaway. While this species is also called the common cuttlefish, that's a terribly uninspired name for such a striking camouflage artist. Their skin pulses with iridescent light, rapidly transforming into multicolored patches that help them blend seamlessly into any background. They have large, curved eyes and pupils shaped like a wavy W, a face full of eight arms, and two spring-loaded tentacles that can shoot out to grab prey. Two of their arms can be raised above their head to mimic horns when they feel threatened.

These patchwork cuttlefish can only be found four or five months of the year in the Seaforest. Nobody knows where they go the rest of the year. While they are here, they share the caves with pyjama sharks, which are a major predator of the octopus yet live harmoniously with the cuttlefish.

Cuttlefish taste and smell (to humans, at least) very similar to octopus, but they're faster and more capable of outmaneuvering a shark. They're also incredibly sensitive and the slightest disturbance will send them jetting away, leaving even the fastest pyjama shark or human swimmer in the dust.

Though Aaron is a fine athlete, he was new to diving and was struggling to swim easily. I led the way, swimming gently toward the cuttlefish, careful not to disturb the water. And sure enough, they didn't move. Everything was perfect, and I was able to bring my lens mere inches from their faces.

With their arms dangling in front of their faces, the cuttlefish call to mind wise old wizards. And they are, indeed, ancient—the deep cephalopod ancestor of these animals evolved over 500 million years ago, long before the first shark appeared in the oceans. Like their cousin the octopus, cuttlefish have large brains and are remarkably intelligent. With their soulful expressions, they regarded me and my camera with great curiosity.

Of course, what they see looks different from what we see.

As far as we know, they should only be able to see in black and white, despite their masterful ability to camouflage themselves in a rainbow of colors. They do this by using their W-shaped pupils to create blurred color around the edges of objects—a phenomenon called "chromatic aberration." Some of my cheap old film lenses used to do the same thing. Their strange pupils allow light to come in from all directions, blurring the image and letting them see color despite being technically colorblind.

They can also distinguish polarized light better than many other animals, adding another dimension to their vision.[1] To us, polarized light appears simply as glare, which we block with sunglasses. To cuttlefish, polarized light is practically a secret language, conveying visual information that helps them communicate, see objects in high contrast, and find prey.

After filming for a few moments, I signaled Aaron to approach.

He eagerly swam over, but as he drew closer, his long fin hit the seafloor and blasted the area with a huge plume of sand.

In a second, the cuttlefish had scattered.

Back on land, Aaron was devastated. But over the years, I've kicked up more plumes of sand and scared away more cephalopods than I can count. It takes time and patience to learn to move differently, to learn a language that is part of us but so removed from modern life as to seem a foreign tongue.

Skating on the Surface of Understanding

When I first returned to the Seaforest, it was like a deep immersion into pure sensation—I had to give myself fully to the elements, open myself up to the mysterious discoveries hiding under every rock.

But a few years into my diving routine, I began to need more than immersion—I wanted to understand. I didn't want to be just a tourist, a sightseer appreciating the restorative power and beauty of nature; I wanted to speak its language. I wanted to learn this wild vocabulary, use it every day to read the earth, the water, the animals, to see inside a world beyond my greatest imaginings.

But I didn't know how.

I grew up as close to nature as seemingly possible, living in our little house below the high-water mark, swimming every day in the intertidal zone. But proximity does not equal fluency. Some of our greatest scientists were never taught the tracking skills so many of our ancestors would have possessed, never had the mentorship and the guidance every wild child would begin to receive not long after learning to walk.

Even my father, plunging into icy waters in his double rugby shirts, did not know this language. And after years of diving into the Seaforest each morning, I still felt like I was skating on the surface of understanding—a deeply frustrating, even painful feeling.

The San trackers I'd spent years filming spoke this language effortlessly, while I heard only a silence inside. If I could only follow the footsteps in my mind I knew they would lead me back to my ancient ancestor who was a master tracker. I could sense him—my original self—there in the deepest depths of my being, trying to teach me what I needed to know.

Our species survived 300,000 years thanks to our ability to understand the behaviors of wild creatures, following them for many miles across harsh terrains, and here I was, sitting at the far end of that lineage, still struggling to string together two words of my ancestor's natural-born vocabulary.

As much as I yearned for this understanding, I worried it wasn't possible, that I'd waited too long in life. I suspect this is something many people feel when they first identify a desire to connect to nature or to learn the ways of our ancestors. Where to begin when there's so much to learn, when modern human life is so cut off from wildness? Can someone who learns the language later in life ever speak with the ease of someone who was taught as a young child?

Wayfinding

Not long ago, the celebrated Native Hawaiian navigator Nainoa Thompson visited me on the shores of False Bay. After a morning of diving, Nainoa, who is the president of the Polynesian Voyaging Society, told me of his own journey developing this deeper connection with the wild.[2] When he first asked master navigator Mau Piailug to teach him how to use nature signs rather than instruments to sail in the open ocean, Mau had reservations. At the time, Nainoa was still a young man, only in his early twenties, but Papa Mau told him, "You're too old. You need to start when you're three."

Eventually, Nainoa convinced Mau and his other mentors that he was dedicated to learning the traditional path of wayfinding. He studied for years, learning about cloud formations and ocean currents, the flight patterns of seabirds, the difference between swells and surface waves. Using the ancient wisdom that was shared with him, he and his crew sailed their traditional double-hulled canoe, the *Hōkūle'a*, thousands of miles across Polynesia and completed a three-year circumnavigation of the globe without the aid of modern navigation instruments.

Stars were the key guiding elements, as were the sun, winds and clouds, seas and swells, birds and fish. The switch from open-sea birds to island birds would tell them they were closer to land. Deep knowledge of the habits of birds—like the way white and noddy terns will fly back to land each evening after feeding—would show them the way.

As they sailed, Papa Mau would lie asleep in the hull of the giant canoe, feeling the swell with his body. If Nainoa navigated just a degree off course, the old man would wake up and redirect the crew. It seemed to me that his whole body had become a wayfaring instrument, a kind of human radar system. He would feel the angle of the swell and how it slapped up against the double-hulled canoe.

This might have sounded unbelievable if I hadn't seen San trackers from the Kalahari doing a very similar thing when tracking animals. At a certain point, they'd stop following the physical tracks and instead just allow their bodies to follow the animal.

Yes, our bodies can operate in the same way as those of homing pigeons or big cats.

We can find our way home.

But when I moved back to the Seaforest, I was twenty years older than Nainoa had been when he began his training with

Papa Mau. Was I now too old to learn the powerful language of tracking? Every time I thought I was getting closer to understanding, *poof*, the words would float out of my grasp, my fumbling fins sending them scurrying away in a cloud of confusion.

Talking with God

I have met so many people who feel a yearning to connect with the natural world and understand its healing wisdom. Our souls crave a deeper relationship with nature, a connection to the wild developed over a million years of evolution. Yet we have traded in nature for this unreal substance called money, and for the idols of power, status, and ease. We have given up the most real and life-affirming nourishment for a fiction that amassing material wealth and depleting natural resources is the key to our well-being.

In an interview with NPR in which he discussed the impending loss of thousands of the world's languages, anthropologist Wade Davis noted that a language is more than just vocabulary and grammar:

> It is a flash of the human spirit, the means by which the soul of each particular culture reaches into the material world. Every language is an old-growth forest of the mind, a watershed of thought, an entire ecosystem of spiritual possibilities.

What are the implications of losing the very first human language, which enabled us to understand and communicate with the wild? What are the implications of losing this connection to our earliest human kin?

As Davis pointed out, languages die not because it's natural for them to do so but rather because of powerful forces of oppression. But that, he argued, can also be cause for optimism.

> It implies that if humans are the agents of cultural destruction, we can also be facilitators of cultural survival.[3]

That's why, in spite of my fears, I had a strong belief that I—that *anyone*—could relearn to speak that wild language by learning tracking. Following signs and tracks, using alarm calls and smells to find animals or avoid predators, is what humanity has done since our beginnings.

When we track we feel alive, we feel present, we feel useful. We reignite something that has blazed inside of us since the beginning of time.

"Tracking is like talking with God," the San tracker !Nqate Xqamxebe told me.

This ancient art is a form of communication with nature that allows us to swim beneath the veil and pluck the stories of wild creatures to the surface.

Dreaming of the Trackers

Long before I ever dreamed of learning to track, I dreamed of the trackers. Damon and I grew up hearing stories of the great trackers, both past and present. We learned of San hunters whose tracking skills were so sophisticated they could recognize the tracks of hundreds of different species, identify every animal's call and cry. The most seemingly insignificant markings on the ground or scratches on the trunks of trees communicated a wealth of information.

San women were expert foragers who knew the uses of thousands of wild plants, mushrooms, and berries, whether edible, medicinal, or mystical, and where to find them at different times of the year. The women were brilliant trackers too, and would sometimes hunt with the men, using their skill at mimicking animal calls to attract prey.

We revered all of this knowledge and wanted to learn as much as we could about it.

In my late teens/early twenties I had a friend named Bowen Boshier, son of the legendary renegade anthropologist Adrian Boshier. The subject of Lyall Watson's book *Lightning Bird*, Adrian had gone to live in the bush at a young age and was able to handle venomous snakes, such as boomslangs and giant black mambas, earning him the nickname Rradinoga, father of snakes. Though a white man, Adrian lived for seven years with the Pedi people in Lebowa and trained to become a ngaka, or traditional doctor, in Limpopo.[4]

Bowen took me exploring in the Cederberg, the mountainous terrain north of Cape Town known for its great rock formations and cave art. We trekked across a wild, rugged expanse of lush valleys and rocky overhangs until we came to a cave filled with hundreds of rust-red handprints likely thousands of years old.

There were other images on the cave walls too, figures of humans and animals painted with a brush by San hunter-gatherers.

The handprints were part of a shamanic ritual, we learned. The artist would press their hand in a special mixture of ochre, animal fat, blood, and plant juices, and place it onto the rock. When San shamans touched the stone, they were trying to access power from the world inside the rock, which is very similar to the world under the water. The shaman would use the power and strength gained from the ritual to help those who needed healing.

AS MY BROTHER AND I LEARNED, SOME SMALL GROUPS IN THE CENtral Kalahari were still using the old tracking ways. That set in motion the making of our film *The Great Dance*—and in many ways set in motion the course of my life.

During my time making that film, I'd look at a paw print and perhaps be able to identify the difference between primate and big cat, whereas a master tracker like !Nqate or Karoha Langwane could tell me how long since the animal had passed, which direction it was going, whether it was wounded, and how much energy it had left. They could assess whether we should continue our pursuit that day or set up camp and wait until morning.

We witnessed many extraordinary feats of tracking, but one of the most powerful moments was when Karoha left his body and went into a trance while tracking a kudu, a spiral-horned antelope that is known as the "grey ghost" for its ability to conceal itself in the bush.

Karoha was pursuing the kudu on foot, running barefoot. It was scorching hot, and he kept tripping and falling in rodent holes as our truck bounced along beside him, filming the hunt. Suddenly, his eyes glazed over and he stopped responding to our questions. In the Central Kalahari, you can't see far because of the dense bushes, but we could see that he wasn't following the tracks anymore.

It was the first time I'd ever seen spirit tracking, and I didn't immediately realize what was happening.

Karoha explained it later. He was no longer following the physical tracks of the kudu but had entered the animal's mind, locking on to it like a radar system. For 95 percent of that foot chase, he had no visual on the kudu, yet he knew exactly where the animal was. We knew this was true because every so often Karoha would flush the kudu from its hiding place, despite not following its tracks. It was transformative to witness this extraordinary human ability to become the animal, to feel it inside one's own body.

BESIDE THE FIRE EACH NIGHT, THE SAN TRACKERS WOULD SPIN great epics from the smallest scratch in the sand seen during the day. This ability to tell the stories of the animals they pursued reflected their respect for their wild kin and for the land that nourished and sustained them. They took only as much as they needed, and with each kill, there was an energy exchange that had to be restored.

I remember Karoha and !Nqate talking about the kudu that Karoha had run down and speared in that grueling four-hour marathon. Naively, I asked if they felt sorry for the kudu. My question seemed very strange to them; they needed the meat to feed their children. Karoha spoke of becoming the animal, the hunter becoming the hunted. They tried to explain to me that they celebrated the taking of life and to do otherwise would be dishonoring.

A High-Security System

My fears that I might never learn to track went deep. I worried that my life was passing by, with me on one side of the camera and what I really craved on the other. I felt despair at how much had been lost in the name of technology and progress, at how brutally colonialism had wiped out generations of Indigenous people, severing them from their wisdom, their land, their ability to survive. The San hunters were among the most gifted human beings I've ever met, but their lives were incredibly hard. The severing of this connection to wildness has been most traumatic for those still in touch with the old ways.

I felt a particular sense of urgency when it came to tracking underwater. The wisdom in those waters was in grave danger of being lost forever if we could not learn to speak the language of the ocean. Two-thirds of our planet is water, after all, making Earth more of a water planet than a land planet.

I also had another, quite practical obstacle to overcome.

How exactly was I supposed to track underwater where animal

tracks are swept away in milliseconds, where sound moves differently, where smell is impossible for humans, and where even on the clearest mornings you can't see farther than a few feet away? In the Seaforest, visibility on an average day is around twenty feet. No farther. It's like walking through a dense forest at dawn, trying to peer through the mist.

Though I'd been diving my whole life, I'd never seen an underwater track.

On land, I could look toward the horizon to get a sense of how high up in the air a bird was. Compare that to floating fifteen feet above a seabed you can't see clearly, or twenty feet under an ocean whose surface is whipped with waves. The ocean is always in motion, which means the would-be underwater tracker is too. Our whole orientation to space changes in the depths.

After twenty-five years of documentary filmmaking, I'd learned to move stealthily on land to disturb as few animals as possible. I had developed my own methods of camouflage (though none quite as extraordinary as the cuttlefish's). So much of filmmaking is about being patient and holding back: let your camera lens go where your body cannot without disturbing whatever it is you're trying to film.

But moving in this way on land is a lot more intuitive: stay low to the ground, move quietly, avoid what my friend, the tracker Jon Young, calls "bird plowing"—steamrolling through the forest and waking every bird around.

Underwater you have to be even more sensitive because you're in a medium that can transmit vibrations very easily. You can feel fish and other creatures around you moving, and the water carries pressure waves that can alert you to the presence of a creature before you see it.

Now multiply the sensitivity of our radar system by thousands, and you begin to comprehend the amazing sensitivity of a creature like the gully shark.

I remember one early dive when I found a sleeping gully. For a long time, it was believed that most sharks had to keep swimming in order to breathe, but there I was, witnessing a gully lying still. I would later learn from shark scientist Dr. James Lea that they are able to do this by pumping water through small openings behind the eyes called spiracles.[5] I watched the slumbering shark for few moments, admiring his dark-grey body spattered with black speckles.

Eager for a better shot, I drew close, but long before my lens captured the shark, news of my presence reached him in the form of vibrations from a little fish I'd startled. The shock wave from the fish rippled through the water, waking the shark and sending him on his way.

In my early days attempting to track, the light shining through the kelp fronds seemed less like the stained-glass majesty it does today and more like a high-security system at an art gallery. All around me were the greatest treasures, the mysterious and sophisticated creatures I so desperately wanted to understand. One wrong move and I would set the entire Seaforest on high alert.

I realized I would have to learn to move differently, swim differently, *think* differently.

The Tracker's Mind Asks Questions

Though I had many questions about what underwater tracking might look like, I suspected that some of the basic principles I'd learned from my mentors would apply. I was eager to try out what I'd been taught, and eager to share this ancient wisdom with friends and family.

Tracking is a language that has always been communal—a means of sharing both wisdom and resources. And sure enough, sharing new discoveries with loved ones and allies has not only been profoundly gratifying, it continually expands my understanding of the wild world. Through tracking together, I forge connections with new friends and see old friendships grow stronger.

One of my favorite tracking partners is Craig Marais, an old rugby teammate from school, now a television journalist. Though we hadn't spoken in many years, Craig and I reconnected when we realized we shared the same desire to deepen our relationship with nature.

I met Craig when we were kids at Bishops, which was one of the first schools in South Africa to have Black, white, and mixed-race kids learning together. I thought I would take him out tracking once or twice, but after we got together and started talking, we couldn't stop: we talked about our school days and rugby, about politics, about race and identity in South Africa. As we swam out into the Seaforest day after day, we bonded as brothers in a way that broke through all of the artificial barriers South Africa has created around people of different races.

Craig's yearning for reconnection was deep, and he was persistent and committed. I'm grateful for that, as we now go out diving together every week.

One morning we were going through the basics of land tracking as we walked along a wild stretch of coastline, headed for a shipwreck. "Do you see that something's been moving here?" I asked.

"Do you mean this track here?"

"Yes. What do you think it is?"

He crouched down to get a closer look. "I'm not sure."

"Look a little bit closer. What does it feel like? What's the size of the animal?" I asked. By now Craig knew that if there was a tiny gap between prints, it was likely a tiny animal. A bigger gap meant a bigger animal. "So, what's your feeling?"

"I think it's a baboon," he said.

"Why do you think that?"

"It seems that's kind of the size of them."

"Ah. Maybe look a little bit closer at the toes. What do they look like?"

He peered closer and shook his head. "Oh no, it doesn't look that much like a baboon, does it?"

"What about the nails?"

"Oh, no nails," he said. "It looks more like fingers."

"So what might that be?"

"The clawless otter!"

"That's it!"

The tracker's mind asks questions, and when we are patient and consistent, the wild responds.

Once you know the tracks, the process of track identification becomes so much clearer, but until then, it's a jumble of confusing scratches and lines. It becomes second nature only through the process of doing it again and again.

On that one short trek, we found signs of porcupine, striped field mice, and three types of antelope: the small steenbok, the large eland, and the rare bontebok.

We also found an unusual mark made by the high tide as it dragged a shaggy kelp head back to the sea, leaving a track in the sand resembling that of a huge octopus. It was the kind of sign that might not be immediately recognizable as a track, as it did not belong to an animal. But I'd come to learn that the tracker not only speaks the language of wild animals but also understands the messages sent by earth and sea.

Geologic Tracking

We climbed up away from the water to a place where the last high tide had reached. When the sea water spills over the vast beach, it carries thousands of pumice stones made of volcanic ash all the way toward the vegetation some two to three hundred yards from the sea. These lightweight stones, born from a distant underwater volcanic eruption, can float along on the ocean's surface for years before being washed up. Very gradually, the pumice becomes waterlogged until it eventually sinks, some varieties taking a year and a half to do so.

In my mind's eye I could imagine the massive spring tide lifting them onto the shore and making this very unusual track.

Since I've always felt a deep connection to the land in this area, I'm better at noticing these kinds of geologic tracks than I am at following animal tracks. A heap of washed-up pumice might not be the kind of thing you would notice—or even think of as a track.

When I see such things washed high ashore, I can feel the energy of the sea that pushed them, the power of that stormy day, to the point where I can almost relive it.

Tracking like this requires a lot of intensity of thought and energy. Each object we encounter inspires many questions: Where did it come from? How much time did it spend in the ocean? How long ago did it wash ashore?

Even human-made objects carry their own stories. Close inspection reveals evidence of animals who have built their homes from glass and steel or used plastic toys as flotation devices.

We can collect so much information by engaging with the wild in this way. A stroll that might have once seemed rather boring becomes intensely exciting when we interpret the language of the land and sea, and what we observe can lead us to feel incredible things.

Curiosity Is the Key

As we continued along the intertidal area, Craig Marais and I saw another strange sight: ostrich tracks leading from the beach toward the rocks. We watched five ostriches risk injury walking over slippery and dangerous boulders.

My tracker's mind asked, *Why?*

Why take such a risk?

Part of the answer came quickly, as we watched them nibbling on something among the boulders.

"It might be the amphipods," I said, referring to the small crustaceans that live in the upper intertidal zone.

Craig pulled out his binoculars. "I don't know," he said. "It looks like they're eating that vegetation over there."

As we drew closer, I found the smoking gun that proved his theory correct: tiny, cropped pieces of vegetation. I tasted a piece and it was quite delicious—like broccoli but not quite as sweet.

But why come to this dangerous spot to forage when higher up in a less dangerous area there was flat fynbos vegetation growing everywhere?

I looked closer and saw, among the rocks, tattered pieces of washed-up kelp. The nub of this tracking mystery suddenly hit me: The kelp was acting as a potent fertilizer, enhancing the taste and nutrient density of these tiny plants—and the clever ostriches knew this. It was worth risking danger to feast on this vegetation instead of settling for the fynbos growing farther up in nutrient-deficient white sand.

Think of it: a giant flightless bird, the largest bird on Earth, being nourished by ingredients grown underwater in a golden kelp forest. The satisfaction of our discovery spurred us on.

The First Track

With my curiosity leading the way as I took my tracking underwater, I began to tune in to my senses.

The wind here is often loud, masking the sound for bird-language tracking, so smell becomes the dominant sense—at least before you dive under. Early humans who lived on this coast likely had highly defined senses of smell, as their hearing would have been diminished by the white noise of crashing waves.

Each morning, I could sense some of the many aromas they might have detected. The strangely crustacean-like smell of decaying pyjama shark. The odor of whale, very different from the smell of seal. The high musk scent of otter. By the coast, the air carries more moisture and more salt, both of which enhance smell. In a dry desert, smell is greatly reduced in comparison.

Underwater, the snorkel may allow the smell of bird guano to enter, but that's the extent of it, as we cannot smell underwater. Here sight is critical, the radar. Underwater, the eyes are looking for tiny signs: sight is the sense that closes the deal.

One morning I was diving in the Seaforest when I noticed faint lines on a rock. I swam closer and realized they were made of particles of sand. I followed the trail of sand until I saw its source: a whelk, a sea snail with a spiraled shell. As the whelk moved along the rock, it had left a little trail of sticky slime that the sand had stuck to.

I was inspecting the marvelous little spiraled shell when the enormity of what I'd discovered truly sank in.

This was a track!

Finally I'd found what had eluded me for so long. After years of hunting, I'd found my first underwater track.

And if this one existed, perhaps others did too.

Subtle Signs

Once I started looking, I began to find hundreds of these underwater tracks, some more faint and subtle but still there. And that's when things became even more exciting.

Next I found another sand-slime track on a pyjama shark and realized a turban snail had climbed over its head while it was sleeping in a cave. Then I found tracks of mollusks that had pushed sand away to form trails on the back of a huge sleeping stingray, telling me how long it had been resting up.

Eventually, I began to see little boreholes in the shells of smaller mollusks I found amid the kelp and along the seabed. At first, they all looked the same. But as my eyes grew sharper, I began to notice subtle differences: the slightly oval tiny one was an octopus's drill, while the bigger rounder one was a whelk's. All these predation marks were basically tracks, each one full of clues about the animal's predators and its habits.

I was hit with a sort of sensory overload—a shock to the blunt mind of the *Homo sapiens* who's not been schooled in the wild, who just glosses over everything. It felt almost as if I'd been wearing blinkers that had prevented me from seeing what was right in front of me.

But the cold and my curiosity were sharpening my mind with each dive. I began recording these tracks, and each time I saw something new, I would take note. And in time, these isolated notes formed patterns in my mind that became understanding.

An Everyday Practice

One morning the water was particularly rough, but I was swimming hard, feeling strong and alive, alone in the big, wild sea. Of course, I was only alone in the human sense, for I was surrounded by creatures, wild siblings, and teachers. When I glanced to my left, I saw a fish stuck to a rock several feet above the waterline.

What the hell was it doing?

My entire being was with that giant clingfish, every ounce of my mind wanting to understand its motives. Like us, this fish has evolved to stay out of water (though only for part of its life). I've observed clingfish breathing out of water for up to five hours—as long as they remain damp. Highly adaptive and strong, they have a slimy camouflaged skin that can change color, as well as a "suction pad" that can hold three hundred times their body weight.

Higher and higher it went, closer and closer to a limpet, those tough mollusks that stick to rocks and are hard to remove by a human without a serious tool. What could this slimy fish hope to do? Then, in a fraction of a second, as a large wave cannoned up the rock, the fish grabbed hold of the limpet in its large teeth and twisted its muscular body about a hundred degrees. The clingfish surfed down the rock and in one gulp of its massive mouth swallowed the limpet, shell and all.

I'd been a witness to the secret behavior of an air-breathing, wave-surfing superpredator! On shore, my body was shaking, not from the cold but from a kind of awakening, a sense of being let back into the world of my ancestors. I was full of wholehearted-ness and boundless energy.

Each day I set out to better understand the fish with the am-phibious soul. I found their lairs in deep cracks in the rocks, and I saw where they laid thousands of tiny eggs on kelp blades and under rocks. Then I found their jawbones in otter scat and in octopus dens: *Aha!* I'd discovered their main predators.

I photographed the silver eyes on the growing young and watched them hatching. I saw their mating rituals—this is when it became obvious to me that they could radically alter their color from dark brown to light yellow and shades in between. I figured out how to tell male from female, by a tiny flap of skin that only the female has. This flap is used to lay her eggs side by side with no gaps.

Gathering these tiny fragments of lives took years. Every day for months and months I visited the crack in the rock where a huge male clingfish lived. He got so used to me, he eventually allowed me to put my camera next to his eye for a spectacular close-up.

After years of moving through the Seaforest like a tourist struggling to understand the road signs, I now knew what things meant, where to find the magic, where to bring my friends to blow their minds. Like the cuttlefish that cuts through the murky water with its polarized vision, finally I could see the sand slug trails, the star-shaped urchin bite marks on kelp, and the grazing tracks of limpets . . . hundreds of subtle clues to the secret world of the Great African Seaforest.

Today when I slip into the green-blue water or walk a wild shore, I feel a deep excitement. *What will I see? What mystery will I solve? What will I learn today?* It gives life instant meaning and purpose. The tracks on the ground and the sounds and smells in the air have unlocked the secret doors of the animal and plant kingdoms. They are the golden keys to these wild lives, but also to the inner corridors of my mind. They open my own wildness and let it flood and cool and soothe the tame corridors that can be very frightening.

Watching for Subtle Changes

Tracking is not just about following animals; it's also about trying to fill in the gaps and create a compelling story from fragments of biological information. Sometimes I go in search of a specific animal. Other times I happen upon something that might look like nothing but turns out to be very interesting—if its secret can be unlocked.

Some tracking mysteries take years to crack, and a good practice is to keep returning to the same place day after day, year after year. Returning to the same places, watching for subtle changes, and continuing to ask questions replenishes my curiosity.

This is how I eventually unlocked the secret world of the reticulated sea star, a bright red-orange starfish named for the honeycomb pattern on its thick skin.

For years, I'd occasionally come across reticulated sea stars whose arms were curved into a pinwheel shape. I asked my biologist friends why the stars were contorting their bodies this way, but no answers came, until one day I was diving out at the cave in front of my house and there it was again—a red-orange sea star curled in a pinwheel.

I almost ignored it because I'd seen it so often, and I'd just about given up trying to understand why it did this. But then something drew me closer. On the edge of one of the arms I noticed something very small and bright yellow, and my heart began to race.

Could it be? Ever so gently, I lifted one of the curled arms, and my brain leapt in excitement. The entire sea star was covering up a mass of babies, and instantly I knew I'd discovered the track of a brooding sea star.

The pinwheel shape now made perfect sense: the mother needed to create a shelter to protect her young, and the only way to do that was to curl her arms against her body to form a rough circular sanctuary.

Over a week or so, I watched the young grow under their mothers. What at first sight had appeared as just amorphous shapes gradually became tiny sea stars.

After about two weeks, the young left the confines of the mothers' protective shields, hung around for a while, then mysteriously vanished. Where did they go? Off to another habitat?

But then another marvelous thing happened. I started to look more closely at this species, and I noticed some of the sea stars were much bigger than the brooding animals, even three times their size, and had a slightly different shape.

In that watery instant I realized I might have discovered a

new species. Now I could take this information to Jannes and he could do the hard science of proving that this was indeed a new species of sea star through genetic testing and by analyzing every detail of its anatomy.

Moving through the wild as a tracker is deeply stimulating. Often an entire day spent tracking is not enough because there is so much to see. When we walk across a landscape or swim through a seascape, at first it may seem like nothing much is present, but if we look with a tracker's eye, stories are unfolding everywhere.

Tracking with Temperature

It's common for humans to be afraid of sharks and other shadowy things lurking below. I also fear sharks, but only in the rare places where I know I'm vulnerable. The key is to understand the ocean floor topography and how sharks use it. I'd also recently begun using water temperature to track sharks. After many years of observing them, I'd learned they favor warm patches—perhaps to aid gestation, though I'm not sure. So when Craig Marais asked me to help him overcome his fear of sharks, I decided we would go in search of the spotted gully shark.

The water that day was very cloudy, and he was jumpy, getting one or two little frights when he brushed up against the kelp. I've gotten quite familiar with gullies over the years, so I was relaxed. With their cat-like eyes and wide mouths jammed with pointy teeth, they look scary, but they are entirely harmless to humans. We were in shallow water thick with kelp, making it almost impossible to run into the more aggressive white shark. False Bay used to be home to a sizable population of great whites, but they have largely abandoned the area due to the presence of orcas and perhaps dwindling food supply caused by overfishing.

I started seeing glimpses of shark bodies just for a second or two before they disappeared from sight. These sharks prefer to cruise around the rocky reefs close to the sandy seafloor. They are remarkably agile and can swim rapidly, their long fins enabling them to turn on a dime when in pursuit of prey. At their biggest, they can get about six feet long and weigh over a hundred pounds.

While Craig was aware that I had spotted the sharks from observing my body language, he didn't startle. He holds his fear well.

After twenty minutes, I drifted into a warm patch of water where I saw the spotted gully sharks appearing and disappearing into the thick mist-like murk, their bodies rippling like pythons with sinewy muscles. It's awe-inspiring to see them close together in a pack. My friend's fear dropped away when he realized they were not aggressive. One brushed past him and he didn't even flinch. The idea of a shark is often much scarier than the animal itself.

Being with them filled us up with a special feeling. It was like being with brothers and sisters from deep time, great-grandmothers and great-grandfathers from 400 million years ago—that's how ancient sharks are. They are older even than trees. Being with these animals is like looking in a very old and tarnished mirror and seeing a reflection of the wild we once were.

Establishing a Connection

After our dives Craig often wants to track a bit longer on land.

After practicing tracking on his own every day for more than a year, he's felt particularly drawn to one animal: the caracal, a nimble cat with long tufted ears that is notoriously hard to track, as it is predominantly nocturnal and highly secretive.

Thanks to his dedicated study, Craig's eyes have sharpened, and even in a vast landscape of shrub he can now recognize signs of the caracal—also called rooikat for its tawny reddish-brown coat. I joined him as he scanned for the slightest flash of red, the barest movement. He was tracking an animal that can leap ten feet into the air and catch a flying bird in its hooked claws. He smiled as he searched for the cat, which is so very hard to find, willing her to make herself known to him.

Though I was chilled and hungry after our morning of diving, Craig reminded me how much it lifted my day the last time we saw a caracal—she had pounced on what looked like a Cape golden mole.

While a proficient hunter, the caracal poses no threat to people and avoids human contact as much as possible. These shy, medium-size cats can survive off tiny animals like birds, lizards, and mice. They are supreme survivors and are still present in places where lion and leopard are long gone, a testament to their stealth and camouflage.

Craig's pupils dilated and he pointed to a spot in the distance. I could only see grass moving in the wind, so he pointed again.

This time, something jolted inside me. Two black-tufted ears protruded from the swaying grass. It was the rooikat!

Craig has established a connection with this animal that allows him to find her in a landscape where she is invisible to almost everyone else. He has put in the time and built the desire and curiosity needed to track this cat who moves like a shadow being. Following a set of subtle visual cues, he has tapped into his natural tracking intuition to focus his attention.

We all possess this intuition, but it takes practice to reignite it.

We watched as the cat lowered her body and belly-walked along the shoreline, massive Atlantic waves breaking over the kelp forest behind her. Was she feeling our gaze and moving away or was she stalking birds? I sometimes find her kills, a pile of feathers and a pair of earthbound wings left on the ground.

As we walked in a curve toward her, I felt the thread between her and Craig grow stronger. Only twenty feet away, the caracal bounded past us, so close we could almost reach out and touch her. Then she vanished into the low scrub.

The joy of being close to this magnificent animal filled our senses to the brim, erasing all thoughts of cold and hunger. We were sated with that age-old exhilaration that comes from tracking and finding the object of desire. I felt fulfilled, two friends out in a wild landscape doing something our human ancestors have done since the beginning of time.

Animal Teachers

What I am learning is that no one teaches tracking better than the animals themselves. If I make mistakes—and of course I still do—the sensitive creatures will flee and I'll pay the price. But when I'm careful, they teach me how to move—gently and quietly with all my muscles relaxed and my movement so smooth I don't trigger those alarm beams.

Sometimes our mistakes are excellent teachers. I am certain that Aaron will never again kick up a plume of sand in Cuttlefish Cave, just as I will always approach a sleeping gully shark with the utmost care. Each animal is sensitive in its own way and will have a different way of teaching you how to move, how to behave, how to think.

Never approach an octopus directly from above. Instead, make your body small, go slow, and send the message that you are not a predator.

Never disturb a pyjama shark on the hunt. The signs are subtle, but detectable: you can tell by the slightly hunched posture, the speed of its swimming, and the way it holds its sensitive nose that it's on the search for prey.

And always give potentially dangerous animals like giant stingrays lots of space. Don't box them in. Keep still and let them come to you.

A Profound Dialogue

One morning I set out for a protected area in False Bay. It was mid-tide, and the kelp was slowly rising and falling in the small swell. Scanning the great kelp bed, I looked for details, my eyes moving slowly over each area.

I saw the flash of a tail disappear underwater and I knew it was an otter. In the dense kelp, these animals move like apparitions, flickering here, vanishing there. For a long time, I found them impossible to follow, but after years of practicing the wild language, I have a much better understanding of where they go when they disappear.

The best way to see otters is from above, but I cannot fly.

Or can I?

I swam out into the bay, watching the kelp gulls closely. They were tracking the otter too. They knew it was hunting and were hoping to steal a scrap of food from a kill. I swam slowly, without arm movements, just gentle low-pressure finning. Then I projected my mind into the air and borrowed the gulls' eyes, looking where the birds were looking, watching their heads and the sight line of their sharp eyes.

After a few moments observing the gulls, I felt I was close enough to look for the next sign. Then I saw it—the bubble trail! Years of observation have taught me the otter's dense fur releases bubbles into the water in a trail. My heart juddered in excitement.

This trail told me the otter was hunting, searching with its dexterous fingers, probing the smallest caves and crevices. I hung back so as not to spook it, marveling at its speed and underwater agility.

It emerged from a narrow crevice clutching a fish—a two-tone fingerfin. My mind was reeling. This is the upside-down fish I discovered years ago. These fish swim normally, but when they enter caves they roll upside down and pin themselves against the ceilings, hiding in the thick algae.

I eventually discovered that this is a technique they use to avoid being eaten, because if they hide the right way up their eyes cannot see danger coming.

Mysteries were merging; the Seaforest was unveiling its secrets.

The otter clambered out onto a rock, immobilized the fish, and began to feed. The gulls hovered above, wheeling and crying as they begged for a piece of the meal. I edged closer, pulling myself up onto the rock. There I sat hunched over, trying to appear small and nonthreatening—not so easy at two hundred twenty pounds: I outweighed the otter sixfold. But the animal tolerated my presence and allowed me to join him for supper.

I was close enough to see that he was male and that his teeth were worn right down. I suspected he was close to the maximum age for his species: fifteen years.

Later, as I shared the footage with Pippa, she pointed out a small crescent-shaped scar on his nose—we then went back through years of footage, and realized we knew this fellow. We'd been tracking this particular otter for years! What a joy it was to know this creature—a joy to have slowly, over the years, worked out how to track a hunting otter in the water.

I knew him through his tracks, which I'd seen countless times. I knew him by the marks he had left. In this way I knew him from a past life—not in the reincarnation sense, but because so much of what I had learned about him came from signs he'd left hours or days earlier.

Tracking requires complete presence in the moment, but it is also a means of storytelling, recordkeeping, a way of keeping the past alive.

In this way, the wild language immortalizes our ancestors and animal kin.

The most profound memory the otter ever left me came in the form of a set of paw prints he created from ink, drawn on the rock in perfection. It took me some time to figure out how he'd created this inky artwork, but as I looked more closely at the prints, I understood. He had killed and eaten a patchwork cuttlefish and gotten cuttle ink all over his paws. As he moved away, he'd left behind perfect ink prints that grew gradually less pronounced until they faded with time.

I was reminded once again that this wild tracking language is the oldest language on Earth, a profound dialogue with the wild god within and without. It is a way to speak with our original mother and to hear her talk back. A way to keep our animal kin close, and to have their secret lives locked in our hearts.

The prints walk in my mind, and through them, this otter will always live inside me, even though he has long passed.

CHAPTER FOUR

LOVE

CLOSE YOUR EYES AND TAKE A DEEP BREATH. AS YOU BREATHE IN and out, take note of what you hear. Maybe it's the sound of cars zooming past, your neighbor mowing the lawn, a plane soaring overhead. Reflect, too, on what you saw before you closed your eyes: a paved street, electric wires slicing your view of the sky, restaurant signs butterfly-wing bright. Even if you are reading these words from a country home deep in the woods, you likely don't have to travel far to witness the tame world rocketing up around you at dizzying speeds.

Reflect for a moment on just how *new* the tame world is, how different it is from what thousands and thousands of your ancestors would have seen when they looked around them, what they would have heard.

Now take a moment to experience the world as they might have. Can you see in your mind's eye this wild world?

Picture a thundering herd of antelope so massive it would take hours for them to pass. Imagine tens of thousands of animals, maybe more, moving across the land in a single mighty crush of energy; feel the plumes of dust they'd kick out, the vibration of their hooves.

Look higher now: imagine a sky filled with great-winged birds. Let your mind reach toward the nearest coast, and see a sea full of fish who have never known a net.

In such a world, wild animals would not be strange to you. They would not be treated as oddities to be gawked and stared at, or casualties of some new real-estate development.

They would be kin you knew as intimately as any other member of your family.

They still can be—not by making them tame but by making ourselves wilder.

Calling the Springbok

As you drive out of the Western Cape and into the Northern Cape, you are struck with a breathtaking feeling of spaciousness. It is a place of huge open skies—a vast, dry, wild landscape largely untouched by human industry. Flat-topped rocky knolls seared black by the sun stretch to the horizon, interspersed with mounds of flowing yellow-white grass. This area has the least number of roads in South Africa, and the night sky is spectacular—it has one of the lowest levels of light pollution in the world.

One day Swati and I made the drive north to explore some rock engravings that archaeologist Janette Deacon had shown me years earlier. We were marveling at an ancient "rock gong" that San shamans would have once used in rain-making ceremonies: a huge dolerite boulder that's been split in half by lightning—when struck, it resonates with some twenty-two different sounds that shamans would have used to call upon the rain animals.[1]

While we were exploring the area, we looked down and discovered a beautiful treasure: the curved black horn of a springbok. This gazelle-like antelope, which got its name because it can spring more than ten feet straight up in the air, is South Africa's national animal and the namesake of the country's rugby team. The springbok follows the rain, giving it a mythic status among the San as a rain animal—one can certainly imagine a shaman using the rock gong to call upon the springbok.

We examined the horn's ring-like ridges that grew smaller as the horn curved upward to its smooth tip. Swati and I took turns running our hands over the grooved surface.

"It must have broken off during a fight," I told her. Fights between male springboks are common during the mating period, when the rams defend their territories and lock horns in sometimes bloody battles.

I looked around at the wide, arid landscape and imagined this scene playing out.

I'd always felt a special connection to this species, ever since reading Lawrence G. Green's 1955 book, *Karoo*, in which he recounts the experience of Gert van der Merwe, who witnessed a springbok migration in the late nineteenth century, when the immense herds still followed the rain freely across a wide-open, unfenced Africa.

Gert and his wife and three children were trekking across the veldt with their sheep and cattle when they spotted a towering plume of dust several miles away on the horizon. Then came the rumbling vibration of hooves drumming the earth.

> The cloud of dust was dense and enormous, and the front rank of the springbok, running faster than galloping horses, could be seen. They were in such numbers that Gert found the sight frightening. He could see a front line of buck at least three miles long, but he could not estimate the depth.

Gert and his family took shelter in their wagon, praying they wouldn't be overrun.

Wrote Green,

> The noise was overwhelming. Countless hooves powdered the surface to fine dust, and everyone found it hard to breathe. Gert's wife, who had been watching the rush with frightened interest, had to draw the blankets over herself and the children. The dust had almost smothered them.[2]

It took over an hour for the springbok stampede to pass through the family's camp, along the way trampling their livestock as well as every other animal in their path—snakes, tortoises, rabbits, even other springboks who had stumbled and been crushed by their own kind.

Occasionally during such migrations giant herds would come down right to the shoreline.[3] The frontrunners might try to stop, but the rest of the herd would keep moving, pushing throngs of springbok into the waves where sharks and other predators would begin to feed.

This was how people once lived, encountering wildness on a scale we can't begin to conceive of today. Less than two centuries ago, the skies were full of birds—it took hours for the giant flocks to pass, blocking out the sun as they flew. The ocean was filled with so many creatures—great shoals of fish, thrashing pods of whales—that the surface of the water roiled and bubbled like a boiling pot.

It hurt to realize I would never see such a sight.

A Yearning for Our Wild Kin

Today true wildness exists only in ever-shrinking pockets and patches. While springboks are still abundant, the immense migratory herds live only in memory, a casualty of spreading human development, overhunting, fences, roads, and agriculture.

The impact of this loss on the human spirit is profound. Humans have evolved alongside animals over hundreds of thousands of years. Our ancestors' knowledge of the landscape and the animals around them was vast and deep.

For many of us, the pain of living lives so removed from our animal kin has manifested as a hunger for an emotional connection to nature. Our psyches are adrift, seeking their kin, because that's what humans have always done, what we've always known. We've had these relationships from our beginnings—wild people had solid relationships with at least a hundred or more species. Today, as more animal populations crash and more species become extinct, we feel the wholeness we crave slipping away.

That human desire to feel wildness again, to be near wild animals, has had damaging consequences: roadside zoos, where animals are confined in cramped cages; the popularity of exotic pets that were never meant to be captive; animal tourism, like elephant and camel rides; circuses where animals are made to perform demeaning tricks. But these experiences leave us empty.

Yet we can forge bonds with wild creatures without owning or enforcing tameness on them. We can find ways to reunite with the wild, not by living like early hunter-gatherers but by recovering our ancestral link to our wild siblings.

Learning to know wild creatures on their own terms, uncaged, untethered, and in their own homes, is enormously healing as well as exhilarating. Spending time with animals in the wild has opened up my heart and made me fall in love with nature all over again.

And whatever we love, we want to protect and care for.

One to One, Eye to Eye

I had always been in love with nature and could feel her heartbeat everywhere. She was like a creature to me, a giant of many parts. But while I craved a deeper connection with nature on a large scale, it was Swati who taught me to connect with animals as individuals, one to one, eye to eye—something she'd been taught as a child.

She grew up with extraordinary mentors and family friends, including her father's best friend, a student of the great philosopher Jiddu Krishnamurti, who would often take her on expeditions into nature. Both he and Swati's father firmly believed in animal intelligence and that every animal is an individual.

One of her strongest childhood memories was the time she was bitten by her neighbor's dachshund, a typically sweet-tempered dog that Swati adored. The bite was serious enough that she had to be taken to the doctor. When she got home, her father took her aside. Gently, he encouraged her to examine her own part in the incident: Might she have done something to provoke the dog? After all, the dachshund must have had a good reason to bite her. He suggested she apologize to the dog.

The story stuck with me, and I watched as the way I interacted with animals began to change. As these interactions deepened, I was changing too. We often equate wildness with ferocity—and of course wild animals are fierce. They have to be.

But for every minute of footage I might capture of an animal in predation mode, there are hundreds of minutes more spent resting, caring for their young, ensuring the survival of their species. At first glance, you might look at this footage and think, *How boring. Nothing is happening!*

But the closer you get to wildness, the more you understand

that there is intelligence in this patience, wisdom in the waiting. By slowing down and paying attention to the moments in between the great battles, my connection to these creatures—and to the mysteries of wildness—began to deepen and bloom.

Oceans Apart

I first met Swati at a documentary film festival in Bristol, England. I was sitting in a large room crowded with people when one woman caught my attention, drawing my eye like a beacon in a dark sea. It's difficult to say why I felt so drawn to her. It wasn't just her looks, though I was certainly not blind to her beauty. She had long, shiny hair, kind eyes, a gold nose ring. Her clothing was casual yet stylish. Something inside me recognized her, as if I had been searching for an answer without even realizing I had a question. When I saw her, it felt as if the whole room went quiet.

When we were introduced later that day and she described the work she was doing with Bengal tigers in the forests of India, some wild part of me whispered that I might have met my soulmate. She was deeply interested in hearing about the San tracking practices, and shared a bit about her own experience with aural tracking in the Indian jungle.

After that initial introduction, we exchanged information, then corresponded between South Africa and India every single day for five months. Then, in early spring, I bought a ticket to visit her.

After a long day of travel, Swati picked me up at the airport in Delhi, then drove us to the two-story duplex where she lived with her parents. Seeing and talking with Swati after all this time apart was electrifying, like the pulsing of a flame that illuminates a cave of wondrous art. The art is always there, even in the dark, but it's the flickering flame that makes the static figures dance.

Though I was tired from the fourteen-hour flight and the time change, our conversation flowed as easily as it had during our first meeting, and over the many days and nights of our texts and calls. I felt the same deep sense of warmth and connection with her parents, and her aunt and uncle. Though we were strangers who had just met, we felt like kin.

A few days later, Swati and I boarded a sleeper train to Ranthambore National Park in Rajasthan. I felt so lucky to have Swati guiding the way—she had made this journey many times and knew the director of the park quite well. As I stretched out to sleep in my narrow bunk in the tiny cabin, I watched the unfamiliar terrain pass by, and felt at home and at peace.

A Divine Glimpse

For centuries, Ranthambore was the royal hunting grounds of the maharajas of Jaipur. Today it is a wildlife reserve with a population of about eighty Bengal tigers as well as leopards, bears, mongoose, and many other species. At the heart of the park are the crumbling ruins of Ranthambore Fort, a complex of royal palaces, Hindu temples, and courtyards now ruled by birds, bats, and monkeys.

The park director, Raghubir Singh Shekhawat, a trim, energetic man with a full moustache, had offered to drive us around the park for the next few days. As our host guided our little Gypsy jeep deep into the forest, Swati explained that the thick green bushes, tangled draped vines, and mossy green growth obscuring the soil would make tracking by sight difficult, though not impossible.

"The best way when you're in a jeep is to look for the pugmarks," Swati explained, referring to the paw prints of the tiger. *Pug* means "foot" in Hindi, and every individual animal species has its own pugmark. From the appearance and direction of the tracks you can predict how long ago the animal passed and which direction it was going.

As we drove deeper into the reserve, the soft diffused light played on the silvery trunks of the dhok trees, while blazing scarlet flowers lit the canopy of the chilla trees—also known as "the flame of the forest." Dust rose in a haze from the red soil while small hillocks rolled into the distance. The red dirt track was lined on either side with ancient banyan trees, their aerial roots forming a dense, woody curtain. We could hear birdsong all around us, and I tried to pick out the calls that sounded familiar: the bright chatter of a flock of parakeets, the *tock-tock-tock* of a woodpecker, and the whistling call of a myna. As the jeep bumped along, Swati scanned the ground intently, searching for signs.

Mr. Shekhawat told us we were in a patch of forest near the lake that was home to India's most famous tigress, Machali, which means "fish" in Hindi, for the marking on her face.

Swati nodded enthusiastically. Over the seven years she had been visiting the park, Swati had forged a strong thread to this magnificent cat, known as the Queen Mother of tigers. Not only did Machali fiercely defend her territory against other tigers, she was a prolific breeder who had filled the park with her cubs. During a drought year when prey was scarce, she had lost two canines in an hours-long battle with a crocodile who tried to invade her territory—a fight Machali won.

Swati explained that the tigress was such an icon that to be in her presence was to experience a *darshan*—a divine glimpse. I looked up the word in the English dictionary and saw that it meant "an opportunity to see a holy person or a deity."

While the stories of Machali sparked my imagination, that day there were no pugmarks to be found on the road. Swati assured me we would still be able to track.

"If you can't see the physical signs," she told me, "you listen to the jungle."

"LISTENING TO THE INDIAN JUNGLE IS ACTUALLY VERY SIMPLE," Swati said. "There are three main sounds that you listen out for if you're tracking a big predator like a tiger."

She pointed up at the tree canopy. "The first is a monkey called a langur. They have a bird's-eye view of what's happening in the jungle, and they have a very specific alarm call. When the langurs alarm, it's a good sign there's a big predator on the move."

I hung on her every word. In my time with the San I had learned how important all the senses are for tracking. What Swati was describing was a kind of "seeing with sound"—each sound like a drop of water falling into a still pond. !Nqate once observed that a leopard had passed by us far out of sight. Later, he showed me the fresh tracks. At first, I thought he had some supernatural ability, until I realized he knew how to listen to the deep ripples of sound made by the bird alarm calls.

The sun was blazing overhead by now, and we all reached for our water flasks and drank noisily. Swati and I both had bandanas tied around our faces to keep out the reddish dust that coats everything here.

After conferring with Mr. Shekhawat, Swati continued her lessons in seeing with sound. Just as she was telling us about two species of deer that are very important to aural tracking—the small nervous chital and the large sambar that only alarms for big cats—we heard a whooping sound.

"That's the chital," she said, getting excited. "But it could just be a false alarm."

But no sooner had she said this than we heard the short chirping bark of a langur monkey.

"I can't believe it," she cried. "This is textbook! Could be a leopard or tiger, but something is on the move."

Mr. Shekhawat had already steered the jeep in the direction of the alarm calls, but now he stepped on the accelerator and the jeep leaped forward, bumping rapidly down the rutted dirt track.

And then we heard a much deeper and more resonant call, that sounded like a cross between a cough and a car horn.

"The sambar," she said, referring to the mighty deer that

weighs three hundred to seven hundred pounds and is not so easily rattled. "It's definitely a tiger."

Swati pointed up ahead at a spot where the road diverged into two paths. "Look to your left when we come to the fork in the road."

As we rounded the bend, the alarm calls grew louder and more intense. When the jeep rolled to a stop, the thick grass parted and a huge female tiger stepped out of the undergrowth.

We gazed at the tiger in stunned silence. It was Machali, the most legendary tiger in the world. Perhaps it was the heightened sensations caused by the auditory tracking, but I felt slightly out of body, the creature half real, like an orange, black, and white tapestry floating in the landscape.

I watched Swati's face as emotions surged through me, joy mingled with disbelief. Swati is always like a light in the dark, but now her spirit was on fire, her eyes and face glowing in the presence of this tiger, to which she was connected by such a powerful thread.

Now I understood the meaning of *darshan*.

A Different Kind of Love

Though we'd come from different continents and dissimilar cultures, Swati and I found each other through a shared love of nature and tracking, and I felt closer to her than any person I'd ever met. Before I left India, I was already thinking about when we would see each other next.

Two weeks after the trip, I called and asked what her plans were.

"Well, I'd like to visit again in June or July," she said.

"No," I said. "I meant for the rest of your life."

We began planning her move to South Africa soon thereafter, as she knew how important it was for me to live in the same place as Tom, and by the end of the year I would propose. She would later tell me she didn't even need to think twice about the decision. I didn't either.

I pondered our connection and shared love of wildness as I began to craft a wedding ring from the springbok horn we'd found in the Northern Cape. I polished the ebony surface until it shone, then wrapped it in silver. Beautiful and enduring, it would be the perfect wild wedding ring. As I held the ring in my hand and inspected my work, the past and the future melded together like the pattern of rings on a springbok horn, twisting and curving into a stampede of memories and dreams.

Into the Flow

Swati has helped me see that when I get too caught up in the "terrible feeling," I sever not only my connection to wildness but also the thread to the world around me, including to the people I love.

One of the first shoots we did together was in Namibia, where we went to film the trance dance of the San for a follow-up to *The Great Dance*. It was thrilling to have her join me, though it also meant adjusting my way of working. Damon and I were so used to filming together by then, we were like a seamless unit. We could communicate without a word. But I hadn't had the time yet to get into this instinctive rhythm with Swati.

The night of the dance, it didn't take long for me to become completely swept up in the power of what we were experiencing. We filmed late into the night as the ceremony intensified. The rhythmic cadence of the women singing and clapping around the fire rose and fell in one unbroken braid of sound. The dancers had rattles tied around their ankles made of moth cocoons filled with tiny pieces of broken ostrich eggshell, and I could feel the vibrations as their feet stomped the earth.

I didn't want to miss a moment of the experience, so each time a video camera would run out of tape, I'd race to replace it. Swati and I quickly fell into a rhythm: I would pull out the full tape, hand it to her, and she'd hand me a fresh one. It was all happening very fast, it was dark, and before us this incredible dance was taking place.

Then, just at the peak of the ceremony, as the shaman entered his trance after hours of dancing, I handed her a tape and it fell out of her hand and onto the sand.

"Oh my God!" I yelled, holding my head in my hands, as I began to panic. That tape held an experience only a handful of human beings had ever witnessed. Had the cassette broken? Had the sand destroyed the tape? Were those irreplaceable moments of history lost forever?

Entirely consumed by my tyrannical muse, all I could think of was that I had been given this extraordinary opportunity and I'd failed. My thoughts raced. My heart felt like it was going to explode.

The horrible spell was broken when Swati picked up the tape

and we realized there was nothing wrong with it. I put in the next tape and resumed shooting.

Days later, and after I'd gotten some sleep, Swati and I talked it over. I was ashamed of the way I'd panicked and I let her know.

She held my gaze for a long moment and didn't reply. Then she said, "If we're going to be together, it cannot be this way."

This may have been the first time I was able to step out of my own experience and see how it affected others when I got into such a state. Now my frenzied behavior was risking driving Swati away. I could tell myself it was creative flow, but I knew there was something uncontrolled about my manic muse, something unhealthy and rigid. This had been my first shoot with Swati, after all. How could I have expected to enjoy the same level of communication with her that Damon and I had developed after twenty years of documentary filmmaking?

But, of course, the problem went much deeper than that.

My frantic style of working—the feverish energetic highs, followed by the inevitable exhausting crash—needed to change. Swati was suggesting that I could find a way to make films and not go mad in the process; that I not allow myself to become so exhausted I'd be unable to cope with something as trivial as a dropped tape.

If I was going to open myself up to this deeper connection to everything around me, including the wild creatures I wanted so desperately to understand, I needed to tame my tyrannical muse. Perhaps the reason my creative energy was so out of balance was *because* of this disconnection with nature.

Swati brought so many new dimensions into my life. She taught me not to take it all so seriously—to let go of my need for absolute perfection, the need to shoot things over and over and over again until I was certain I had just the right shot. With her encouragement, and by embracing stability in my work and in my life, I was gradually able to let go of those grueling patterns that had ruled me for so long. And when I let go of that old way of shooting, I discovered I could actually do better work. By shooting less, I could be in the flow more.

This helped to balance the scientist-researcher in me that was always looking for answers, allowing me to be more present with the mysteries of nature, and its many joys.

Once, while we were sitting down to lunch after a day of diving, an enterprising baboon snuck up behind us and took off with our carefully packed container of dates, cheese, and biscuits. While I leapt to my feet to chase after the thief, Swati was reduced to fits of laughter at being outsmarted. I felt foolish for having been angry even for a moment.

We'd come to a similar love of nature via different paths—for me, I fell in love with the ecosystem first, starting with my dives into the wild cold of the Seaforest with my mother and father. Next, my desire to track took hold. Learning the wild language helped me become acquainted with the kelp forest and all of its inhabitants. But after observing how Swati was with animals, my love for each individual life began to blossom.

It was a love I was eager to share with my fellow trackers.

Hide-and-Seek

Jannes and I were diving together when we spotted a fish we had never seen before, with a chunky body, fleshy lips, and prominent eyes. Every day for a week we found our mystery fish in roughly the same area.

At first, the fish was very elusive and wouldn't let us near enough to get a picture. Slowly, as she began to grow more used to us, she allowed us to swim closer. Finally, Jannes identified her as a juvenile black musselcracker, one of the most iconic fish in South Africa. This fish has all but disappeared in False Bay due to overfishing, so we were terribly excited, and we set out to learn as much as we could about her.

Musselcrackers are all born female and change into males much later in life. My imagination tracked this young female back to her parents, who were probably very large—these fish grow to nearly five feet in length, weigh about seventy-five pounds, and can be as old as forty-five years. As their name suggests, their powerful jaws can crush hard-shelled prey like urchins, bivalves, crabs, and sea stars. Could adult musselcrackers be in False Bay? I hoped so, but the larvae could also have drifted down to us, courtesy of the swift and powerful Agulhas Current.

Gradually, the clever fish realized we were no threat and allowed us close to get our precious images. One day we watched in awe as she pivoted her body vertically, aiming at a spot in the sand. Nothing was visible. Suddenly she plunged forward, burying her mouth almost up to her eyes in the sand.

Out she came, spitting out sand, her jaws chomping on the unmistakable multiple strands of a tangleworm.

OVER THE WEEKS, JANNES DEVELOPED AN EXTRAORDINARY BOND with the musselcracker. He started to look forward to visiting her every day. While she was curious about him, at first she stayed at a safe distance. She would look at him from afar, but when he would swim closer, she would hide.

He observed her feeding in the sand, and noticed how she liked to hide in the shadows, how she could swim both backward and forward. As he studied her behavior and tracked her favorite places to hide from predators, slowly, very slowly, the musselcracker led Jannes back to his own ancestral connection to nature.

As a result of all the time Jannes spent with her, he felt his strength in the water increasing, his connection to nature slowly reawakening. He began to alter his swimming to become gentler, learning how to move so he would not scare her away, how to breathe so he wouldn't produce bubbles underwater.

Over the days and weeks, the musselcracker would draw Jannes down to the water again and again.

When he came home from his dives, he'd look through the pictures he had taken of her, analyzing her behavior and looking things up in the scientific literature. Even on a day when he was exhausted from other things in his life, he would walk down to the water in the late afternoon to spend another hour with her, marveling at her clever game of hide-and-seek.

Jannes began to notice a subtle energy shift that profoundly changed the entire relationship. He no longer had to look for her when he swam over the patch of kelp forest where she lived. Instead, she was looking for him! She would come right up to him and then follow him through the kelp.

Jannes told me that being with the musselcracker gave him an indescribable feeling of love and appreciation for each moment. "Whenever I am with her, my senses are heightened," he said. "There is no room for anything else in these moments."

An Ancient Bond

Why would a fish like this musselcracker—or any wild being—tolerate the presence of a human? Perhaps they recognize in us the same thing we see in them: kinship. I remembered how the Nile crocodile, one of the most formidable predators on the planet, had accepted our presence in its underwater chamber, resting on the river bottom as we filmed it. The Cape clawless otter I'd met during my cold-exposure journey—normally an extremely wary animal—had not only tolerated my company but also seemed to welcome it. Machali, the wild tiger matriarch of Ranthambore, was renowned for her relaxed attitude around humans, practically posing for photographs and allowing people to study her behaviors up close.

How can we explain these magical encounters except through

our shared history? Perhaps wild animals, in some deep part of themselves, remember us as we once were—nomadic, wild, free. As humans were for generations before we became tame.

The San trackers I met in the Kalahari were keenly aware of this ancient bond between human and animal. I had watched Karoha go into a kind of trance state during which there was a merging of minds, a sense of oneness with the animal that allowed him not only to track but also to take life in a way that honored the sacrifice of the animal and the exchange of energy. I marveled at the ability, and wondered what Karoha experienced when he went into this state.

Then one day I got closer to understanding.

While filming, we met a young farmer in the middle of the Karoo, a very isolated part of South Africa, who had raised an orphaned springbok from birth. Now grown, the animal had mostly returned to the wild but would still come close enough to the farmer that he could almost embrace it.

Because of this trusting relationship, we had been able to strap a tiny camera onto the springbok. Damon's wife, Lauren, an occupational therapist, had constructed a special harness for the camera that gave us wide-angle footage of the animal's legs and hooves hitting the ground as it ran and leaped about. But the truly extraordinary thing is what happened when we showed the video to the San trackers.

"That's what we see," Karoha cried, pointing at the image of the springbok's hooves striking the dirt. "That's what we see when we go into the animal's mind."

San rock art is full of imagery that suggests the feeling of taking on an animal body as well as being inside an animal mind. The artwork hints at a consciousness in which animals played a powerful role, and not just as a source of food. San depictions of therianthropes—those figures with human bodies and the heads of animals like antelope, elephants, baboons, leopards, and birds—may represent the shaman's ability to transform into an animal while in a trance.

The beautiful images inspired me to ponder what it meant to have a wild mind, to contemplate the differences and similarities between the human and animal minds, and to appreciate what animals have to teach us about our inner states of being.

I shared my musings with my friends, and one day after an icy swim, Pippa hesitantly shared that she'd once experienced the feeling of being in a southern rock agama's consciousness. This is the famous blue-headed lizard endemic to the Western Cape of South Africa. Only the male agamas have the bright turquoise coloring that appears strongly during mating season. They display their prowess by doing push-ups to impress the grey-brown females. As her mind fused with the lizard's, Pippa felt completely present, emptied of memories of the past or worries about the future, and with a heightened awareness of predators like snakes and eagles. It was a liberating feeling.

Her dark eyes met mine as she said, "I actually felt myself doing the push-ups, Craig!" Pippa is a serious and thoughtful person who chooses her words carefully. I knew this was an authentic experience.

Animal Intelligence

Years of tracking, observing, and photographing wild creatures has shown me that just as every human is unique, so is every animal. With each interaction we stimulate each other's lives. While some species of reptiles or amphibians may at first appear to be one-dimensional, all animals are actually very complex, and intelligence is not a hierarchy, with humans at the top of the ladder and basic bacteria at the bottom.

I need only think of one of my favorite animals, the octopus, to understand the complexity of animal intelligence. If we compare octopus intelligence to human intelligence, the octopus may seem way down on the scale, until we look at its marvelous locomotion and its decentralized nervous system: two-thirds of its cognition takes place outside the brain, in its eight arms, which can all move independently.

Imagine trying to operate two thousand fingers like an octopus's two thousand suckers, each with its own sense of taste and powerful grip.

Imagine trying to match your skin color and brightness to your surroundings.

Imagine trying to do blind geometry while drilling into twenty-five different mollusk species.

Beyond that, every individual within each species has its own unique personality. I've met hundreds of octopuses, and each seems different, from outgoing and friendly to shy and fearful.

And, of course, mischievous.

One morning Swati and I were swimming in the kelp forest. It was low tide and she was wearing her wetsuit. As we scanned the Seaforest, we noticed we were not alone—a small octopus was observing us with curiosity. While this was years before I would forge a bond with an octopus that would change our lives forever, we both had a great appreciation for these clever creatures.

I watched as it clamped onto Swati's foot. Then it let go and reached for her arm. I asked Swati to keep still, as it wrapped itself around her arm, then her waist, and back to her arm in a kind of dance. She stood completely still, letting the octopus conduct its investigation, and after about five minutes, it swam away.

It was only after the curious octopus was far away that Swati noticed it hadn't gotten away empty-handed. It had taken her wedding ring, the one I'd carved so painstakingly from the springbok horn. After our initial dismay over the loss, we had a good laugh. After all, it seemed a fair exchange for all that the sea had given us.

Minds, Personalities, Feelings

I would learn much about the unique personality of individual animals from our house cats. To be honest, I was not that keen on the idea of getting cats, or any other domestic pet, but I knew it would make Swati happy. So I didn't pay much attention to the cats, and they didn't have much impact on me—until Leon.

Leon was a large and powerful cat, but also a wonderful mix of brave and sweet. When I used to get the chest infections that plagued me after my bout with malaria, Leon would lie on my chest and purr. If anyone in the house was sick, he would immediately go to them. He always knew when someone needed comfort.

One day we were in a terrific hurry to leave the house, and I had to get him from underneath the bed and put him in the other room with his food and water. I wriggled under the bed as far as I could, but Leon retreated farther. So I grabbed the only part of him I could reach—his front legs near his shoulders—and pulled him toward me. He hissed at me, and in that moment I knew he could have ripped me to shreds with his teeth and claws. But he allowed me to lift him and carry him into the other room.

It was so moving to see that restraint from a big, strong animal. He was so forgiving.

Though I'd never expected to forge this bond, the connection I had with Leon reminded me that domesticated animals still possess wildness, just like you or me. For all her decades of work in the wild, my friend Jane Goodall has said that one of her greatest teachers was her childhood dog Rusty, of whom she wrote: "He proved to me that animals have minds, personalities, and feelings."[4]

And, of course, animals have their own life experiences that imprint on top of inborn personalities to alter their overall behavior and attitudes. Not long after Swati and I built our natural swimming pool, I got to know a Cape river frog who had moved in and made a home among the rushes and reeds we'd planted. I suspect he hitchhiked in on a bird foot as a frog egg.

This large frog is a favorite prey of wading birds like herons, and normally leaps into the water at the slightest disturbance.

Sitting on the edge of the pond with the frog, I felt privileged to be allowed to share his amphibious world. When I dipped in the pond, he would sometimes swim very close to me, unafraid. I loved hearing his rasping croak in the evenings, when he was looking for a mate to share his pond.

One day I came home and found the frog lying on his back next to our cat's bowl, nearly lifeless. Horrified, I picked up his unmoving body and put him back in the water. By some miracle he survived, but his whole personality had changed. He carried the trauma of the cat attack and would not come near me anymore.

Falling in Love with a Fly

It's one thing to fall in love with creatures we consider regal, clever, or sweet, but what about those we've learned to swat away?

I've never liked flies, especially after an encounter years earlier on the great Gariep River, which stretches from South Africa into Namibia. At the time I had begun experimenting with creating land art, which I'd felt drawn to after seeing the work of Andy Goldsworthy. I struggled greatly at first before managing to find my own simple style—placing natural objects like bones, stones, shells, and kelp in patterns on the ground and then photographing them in the landscape.

Not long after I'd arrived in the area, millions of tiny flies that had recently reproduced were congregating in giant swarms. Most of the other people fled the area until the swarms passed, but I was committed to venturing out, desperate to do my art. I was bitten everywhere, and no amount of waving a branch over my head helped.

Eventually I tied a mosquito net over my head. Still, it was very hard to tolerate the swarms, which sometimes dimmed the sun. The only positive was that I had the entire wild landscape to myself.

And so I couldn't imagine loving a fly. But that all changed when I got a glimpse inside their lives. It first started when Jannes and I were out one afternoon and I noticed strange little tracks on bare granite rock in the upper intertidal zone of the ocean, the area that gets wet at high tide only. I asked Jannes about these tracks, which looked like little white squiggles, and he thought they might be traces of salt left in the cracks. We kept walking, and then I saw them again.

By that point, I'd been focused heavily on tracking and my eyes were keyed to pattern recognition. Something told me these weren't random salt traces. Jannes and I got down on our hands and knees and looked up close.

There we found strange little leathery casings that had attracted tiny particles of sand. Suddenly Jannes gasped. "Oh my God," he said. "It's larvae."

Totally hidden on what looked like rock around us were hundreds of animals. We had no idea what they were because they were still in larval stage.

"A beetle, maybe?" Jannes wondered.

I had no idea. We sent the pictures to University of Cape Town professor Charles Griffiths, who had become my trusted mentor in all things marine biology. He thought it was the larvae of a marine fly.

"Marine insects are rather rare," he added. "Please try to find the creature that hatches once the larvae transform, as it could be new to science!"

This is what I love about tracking: I get to become a nature detective.

So every day I kept returning to the same spot. With Pippa's help, I found empty casings where the animal had hatched—and she found one rather strange looking fly still attached to its cocoon with its wings all curled and folded up.

I waited for movement, but there was none that day. I did discover many of the larvae were actually underwater in shallow pools—this truly was a marine insect.

It was when I began to do further research on this humble fly's life cycle that I discovered something fascinating. The flies lay eggs, which then hatch into larvae. These larvae are eating machines, the great composters of nature, turning dead matter into nutrients. But the really interesting part is when the creature has reached the end of its larval cycle but before it becomes an adult fly. I've always thought the larva somehow grows wings and legs and transforms that way, but what happens is very different.

The larva first turns into a liquid, and somehow that liquid knows how to make a creature that can defy gravity and fly. Out of the liquid, the atoms rearrange themselves to produce a perfect little flying machine.

If you gave a bunch of our best scientists a billion dollars they'd be hard-pressed to make a fly from scratch.

Nature is priceless.

A week later Jannes and I set out to find our mystery fly. It was midwinter, the wind was freezing, and I was not confident about our chances. We searched everywhere, and just as we were about to give up, in the last pool we found one perfect fly resting on kelp. I had a few seconds to set my camera to macro and take a few images before the wind blew the fly away.

How does an animal made of material softer than paper survive and thrive in a place that is so hostile to something so small—where huge waves thrash against rocks, temperatures fluctuate wildly, and predators abound?

After just a week of tracking that long-legged silvery fly, I glimpsed its magnificence and its magic.

I'd thought it impossible, but I'd fallen in love with a fly.

The Gentle Push

The sea may seem like a mysterious and strange place, inhabited by creatures who look and act so different from us. Yet the more I spend time in the ocean, the more I realize all creatures are manifestations of life—each a piece of the shining totality that is nature, or God, if you like. All living things on Earth—humans, dogs, and frogs, sea cucumbers and musselcrackers, bananas and banyan trees, and, yes, even the common fly—are descended from a common ancestor that came from the sea. We are all made of the same substance, animated by the same life force.

Falling in love with the living planet around us—surrendering to her fully—is a good way to begin to know her primordial character. We can't force serendipitous insights and discoveries, but we can keep our curiosity bubbling, and often it's the gentle push that opens the gate. That's happened to me often enough that I can't deny it. Experiencing this mystery deepens my love for the natural world.

Understanding the subtle behaviors of our animal brothers and sisters fills me with meaning and wonder. It feeds my primal mind with a food that nourishes a deep evolutionary drive. When you have an intimate knowledge of other species, you also know how to find food, which gives you self-sufficiency, a sense of place, of belonging. If you don't know the species in your own backyard, you don't belong to your own birth world.

Nature is our first mother and we are inseparable from her. She gave birth to us as surely as our human mother did, and we are woven into her, made of her bone and her blood. We take her in when we breathe, when we drink, when we eat. She has nurtured our most distant ancestors and our animal kin, and every second of our lives she has nurtured us.

As I get to know the Seaforest more intimately, I feel my body and mind merging with her, falling in love with her at deeper levels. I can feel weightless tendrils unfurling from my being and reaching into the wilderness. I feel them pulsing with life and the teachings of creatures and place. Big Mother is speaking my name—not the name I know but another name that is ancient and mysterious and hard to comprehend. I must listen carefully so I can hear the whisperings and follow their threads.

CHAPTER FIVE

ANCESTRY

ONE MORNING I WAS KAYAKING ALONG A WILD SECTION OF COAST-line where the steep cliffs are pocked with sea caves. It's a beautiful place, but dangerous too. Most days the wind and swell are much too powerful for kayaking or swimming, and the water is so deep that whales and sharks swim right up near the cliff face.

I was closer to shore than usual, and as I was about to paddle around a promontory, I spotted a cave I'd never noticed before.

The waters were calm enough to allow me to catch a little wave and surf the kayak right onto the rocky shelf in front of the cave. I leaped out and dragged the kayak higher up the rocks. Then I ducked through the cave entrance, which was just wide enough for me to fit through comfortably. Sand and shells crunched underfoot, and massive piles of shellfish lined the cave walls.

A stream of light poured in through the entrance, and I began to examine the shells, wondering what stories they would tell me about the animals who lived here. As I knelt to get a closer look, I recognized the unmistakable shape of a human skull sticking out from among the mollusks and seal bones.

It's a strange thing to be confronted with a human skeleton. All your senses are suddenly on high alert. And so I approached the bones slowly, careful not to disturb anything. I could see the cranium and the lower jaw in piles a few feet apart. When I looked closer, I spotted human femur and rib bones.

At first glance, it seemed very recent—possibly a fisherman who'd drowned. But then I looked closer. The teeth were worn flat in the classic way of all hunter-gatherers who eat food gritty with sand and prepare skins by chewing them soft. There was not a single cavity on any tooth, attesting to a diet without sugar.

This person had lived fully wild.

The bones showed no evidence of the cause of death. I guessed that the skeleton was probably five hundred to a thousand years old, maybe even older. But the bones were remarkably well preserved, thanks to the lime from the shellfish.

Around the body lay evidence of the person's last few meals: mounds of shellfish and seal bones. I was also struck by just how difficult it was to get to the cave. This person must have been a powerful swimmer and diver.

As I stared into the eye sockets of the skull, I thought about what this person's life would have been like. Our ancestors saw things we could not imagine, nature at her most formidable. For hundreds of thousands of years, they built a deep relationship with the sea and land and their respective creatures—the wilderness gave them life, sustained it, and took it away, and then the cycle would begin again.

It hit me like a tidal wave—five thousand generations of people had lived in this place. Like most things in archaeology, it's impossible to say for sure, but as far as I know, in no other region on Earth have humans had such a long, unbroken relationship with the coast and the sea. I began to tremble from the realization that somehow I had been brought into contact with this magnificent ancestor and here I was, looking into the end of this lineage of wildness. I backed away without touching anything. I felt strongly that I should leave the bones as I had found them, that the person should stay undisturbed.

How I wished I could spend just one day with that person. I'd probably learn more in that day than I had in my entire life.

But these ancestors were gone, and with them went generations of wild wisdom.

Was there a way to reestablish connection?

Broken Threads

As I deepened my tracking practice, the pieces of a puzzle that had been floating in my mind began to fit together, and the picture that was emerging linked me back to the earliest humans. I could feel part of myself lighting fires in caves thousands of years ago, foraging on land and sea. I started to feel part of this place, embedded in nature.

But despite everything I was learning about wildness in the here and now, there was still so much I didn't know about the people who walked this coast for thousands of years, the fearless ones who first dove into the waters that had been healing me since I was a child. Had I been born here a thousand years earlier, the knowledge these ancestors possessed would have been passed down to me, but that connection has been severed, and so much wisdom lost.

The San refer to this connection to our ancestors as "threads." While I was filming *Cosmic Africa* in northeastern Namibia, I met the shamans Kxao Tami and |Kunta Boo. During rituals that brought them into heightened states of consciousness, they could actually perceive these threads. |Kunta Boo described the experience as akin to climbing a glowing spiderweb to his ancestors in a bright and shining place.

Therapist and author Bradford Keeney spent years with this group in Namibia and later described his experiences in his book, *Ropes to God*. The shamans generously initiated Keeney in their healing ceremonies, and he entered trance states that allowed him to perceive at least some of this connection.

Some people have never lost this connection. But the majority of humans today—particularly those who are not living on their ancestral lands or in their ancestral communities—have a growing hunger to mend those broken threads. I've witnessed this desire in people of all ages and from many different backgrounds who've asked me to help them reconnect with the wild.

Take my young friend Gaz, a marine biologist who was educated in a Western-style schooling system in South Africa. He described a sense of feeling lost, disconnected from the religion of his parents, hungering for deeper meaning. What was his place in this world? How did he fit into the big picture? He was trying to find a spiritual dimension to his life that made sense to someone also trained as a scientist.

Things started to fall into place for Gaz when he began to learn more about our wild ancestors, and the way humans lived for the majority of our time on the planet.

We were tracking along the coastline when we discovered a Stone Age shell midden, a site where early people ate and discarded shellfish and animal bones. I showed him an eland bone that had been split open so the marrow could be scooped out and eaten. There among the heaps of ancient shells and bones, Gaz found three ostrich eggshell beads from a necklace that was possibly several thousand years old.

Ostrich eggshell beads are among the oldest known human-made ornaments. These tiny white discs had probably been worn by different people over several lifetimes. Holding in his hand a bead worn by a fully wild person only a few thousand years ago, the connection became real to Gaz.

This was not in a book or a film or a museum—we were standing there in the crucible that made us human beings. We were trying to solve the same tracking mysteries that these early humans had worked out. We could smell the same rotting kelp, salt air, and metallic ozone that they had smelled. We were diving in the same ocean. And suddenly Gaz didn't feel so lost anymore. He had something to hold on to. He could feel his ancestors walking with him.

We lit a little fire on the beach and roasted our food over the flames, telling stories and jokes, and there we became part of the greater human story.

By understanding a little about our ancestors—their lives, their technology, and their connection with other living things—we can better understand who we are as a species and where we are headed.

The First Amphibious Souls

With thoughts of our ancestors in mind, I left the cave where I'd found the skeleton and dove into the water, enjoying the feeling of zero weight on my body. Swimming and diving is such a primal activity—our amphibious origins go deep into prehistory.

Neanderthals were very likely swimming on the shores of Italy about 90,000 years ago.[1] Archaeologists found evidence that they could have dived down to depths of up to twelve feet to collect smooth clams for eating and for making tools in their caves. Aural exostosis, or surfer's ear—the condition I have in both of my ears—was found in Neanderthal skulls, suggesting they spent a significant amount of time diving and swimming in cold water.

Modern humans with European ancestry only have about 2 to 4 percent Neanderthal genes, so I don't feel overly close to these sophisticated beings who once roamed Europe but became extinct about 40,000 years ago. I feel much closer to my *Homo sapiens* ancestors who once lived here at the southern tip of Africa. Evidence for use of seafood by our species goes back well over 100,000 years, and I've found hundreds of middens along our coast laden with the remains of the kinds of shellfish they would have enjoyed.

If Neanderthals were diving in Italy 90,000 years ago, it's very possible that our species began diving in the relatively safe kelp forests of Africa more than 100,000 years ago. At around 100,000 years we see a sharp increase in cognition. The archaeological artifacts from this period show that as a species we started to wake up in a special way around this time as our thinking slowly began to evolve.

The deepening of our relationship with the sea perhaps happened somewhere between 100,000 and 120,000 years ago, a period when humans became curious and highly inventive. The water would have presented a frightening but irresistible pull to such an inquisitive mind.

There are also very practical reasons for taking to the water. Food would have been a major motivation for swimming and diving, but the ability to cross rivers and estuaries also would have made life much easier. Being able to swim would have been a great survival advantage if a large wave snatched a person while gathering seafood from the shore—a not infrequent mishap.

And, of course, there is nothing better than a cool dip on a hot day and the sheer joy of being in water. Early people had lots of time to play, learn, and experiment, as they needed only a few hours a day to meet all their food and shelter needs. It's hard to believe that this huge human brain would not have watched animals swimming and realized that this was something very worthwhile. On very hot days, I've seen baboons swimming and diving in the kelp forest, some as deep as five feet underwater. The shallow rock pools, protected kelp forests, and estuaries of the South African coast with its mild climate would have been the perfect place for humans to first learn to float, to swim, and eventually to dive.

The World's Oldest Time Capsule

In an effort to better understand our wild origins, I sought out the wisdom of South African archaeologist Christopher Henshilwood, who made history in the 1990s by unearthing remarkably preserved evidence of early human behavior at Blombos Cave on the Southern Cape. Christopher, whose family has lived in South Africa for several generations, and his partner, archaeologist Karen van Niekerk, have found the oldest known abstract drawing—dated to 73,000 years—along with the oldest paint container—100,000 years old—made from an abalone shell, and the oldest pressure-flaked stone tools. In essence, they've discovered that our prehistoric ancestors were innovating at a very sophisticated level between 60,000 and 100,000 years ago.[2]

I was initially quite nervous to reach out to Christopher. He's six foot six with a deep baritone, but it's his keen intelligence and extraordinary discoveries about modern human origins that make him such a formidable presence. When he saw my dedication to studying our wild ancestors, we became friends, and he started to mentor me one-on-one as I began to create films and exhibitions around his discoveries along with Damon and the archaeologist Petro Keene.

One afternoon Christopher invited me into the inner sanc-
tum itself, Blombos Cave, which is like going to the mecca of
the human species. The cave is off-limits to the public, accessi-
ble only for scientific study, and visitors have to be exceptionally
careful not to disturb anything.

We trekked through a nature preserve to the coastal cliffs
where the cave is located. The landscape was rugged, and it was
a steep climb to the entrance, which is marked by a nearby giant
rock arch.

Inside the cave, it was quiet—the thick rock walls muted the
sound of the wind—and though it was summer and quite warm
out, the temperature stayed cool. Around us, scientists were
working quietly, almost with a sense of reverence. Researchers in
masks and gloves were bagging artifacts and entering information into digital tablets. One scanned the cave walls with a laser
to record the exact location of each object to determine the time
periods of each layer of discoveries.

I watched where I stepped, mindful that if I slipped and put
a hand out to break my fall I could destroy a piece of human
prehistory. This was hallowed ground, a sort of archaeological
Library of Alexandria. This place held the key to everything.

Christopher walked me over to one of the cave walls. He
wore a pair of magnifying goggles, like a dentist, and had a set
of tiny instruments that allowed him to slowly and painstakingly
brush away a layer of rock to reveal what was underneath without harming it. He and his team had spent thirty years carefully
scraping away layers of the rock surface, peeling back time so

we could see into the past and unlock the greatest secrets of our species.

The early humans who used the cave were highly nomadic, he told me.

"Each time a group of people came in the cave they would only spend a few weeks and then keep moving," he said. "Then the sand would blow in and cover and perfectly preserve what they left until the next group arrived months or even a year later."

About 70,000 years ago, the cave filled up with dune sand, basically sealing it off like a time capsule. The sand prevented oxygen from wearing away at the impressions on the stone and the artifacts embedded in each layer. So as each object is excavated, it's as if it were dropped there only days before.

"Look here," Christopher said. "This knapped stone looks like it was worked yesterday. And this turbo shell is a hundred thousand years old, but it only looks a few years old."

As he showed me an abalone shell embedded in the one-hundred-thousand-year-old layer of earth, perfectly preserved by the extraordinary conditions in the cave, I realized I was looking at a real-life time machine. These findings offered an intimate glimpse into the lives of our Middle Stone Age ancestors, even illuminating the ways they thought.

The cave also contained evidence of the oldest chemistry on Earth: two paint-filled abalone shells, tool kits for painting that held a mixture of red ochre, charcoal, and ground bone. It struck me in that moment just how sophisticated we were so long ago, and although it's hard to prove scientifically, it screams ritual. Stone Age cultures across the world have used this combination of red, white, and black in their ceremonies, and a very similar mixture with the same colors was found inside a pyramid in Egypt. Even today advertising agencies know these three bold colors make the strongest impact on the human psyche.

Christopher and Karen's finds told the story of how our species first began to record information outside of the human mind—how we began to use symbols. The collection from Blombos Cave included a rock fragment with a cross-hatched pattern sketched with an ochre pencil—the oldest known abstract drawing made by human hands.[3] Together with all the other supporting evidence, it was in some ways akin to the very first computer or book. This was the beginning of human genius, a symbolic language that could store accumulated data and pass it from generation to generation. These discoveries completely altered our understanding of human prehistory. Where archaeologists once believed Europe was the hub of modern human cognition, it is now generally believed that the most creative innovations took place in southern Africa.

What I saw in Blombos was enough to feed my amphibious soul for many weeks. But it was a discovery that Christopher and Karen found not too far away, in Klipdrift Cave in the De Hoop Nature Reserve, that would deepen my understanding of wildness—a human advancement that threatened to sever our ties with our wild kin, but that perhaps also held the promise of reconnection.

The Birth of the Bow and Arrow

I was holding in my gloved hands one of the world's most precious artifacts. Sealed in three consecutive plastic containers, swathed in special acid-free papers and cocooned in bubble wrap, unwrapping each box was a bit like unnesting a set of Russian matryoshka dolls.

I nearly gasped as I removed the final layer to reveal the objects inside: two crescent-shaped, beautifully knapped stones—the oldest arrowheads yet found, dated from approximately 66,000 years ago.

The technology of this weapon was remarkable. The head of the arrow was two little stone crescents mounted back-to-back with a glue-like mastic. Upon impact, the two pieces would come apart, and the arrowhead would open up like a dumdum bullet that expands upon impact. Similar-aged arrowheads were found in other caves along the South African coast, showing that people moved great distances and shared technology.

Consider what the creation of projectile weaponry did to our relationship with animals, and our place in the wild.

Shooting from a distance changes the whole dynamic between hunter and prey. Imagine a big predator like a lion or a leopard: taking on a human, even one with a spear in hand, wouldn't be that risky. The big cat would have a fighting chance of coming out the victor in that contest.

But once humans came up with the bow and arrow, that lion or leopard would need to spot tiny projectiles flying at them from fifty yards away. And coating the tips of arrows with poison gave humans an even bigger advantage.

The human being became lethal.

Armed with weapons that allowed us to kill from a distance, humans vaulted to the top of the food chain, becoming the planet's most fearsome predator.

My time with the San brought a deeper dimension to my understanding of this technological advancement, and of the power we have over the tools we create and use. I've been fortunate enough to hunt with Xhloase, the bowhunter. He showed me how he prepared his arrows, which were made from fencing wire pounded flat on a rock and then fitted via a special wood collar to a reed shaft. San bows may look delicate and lightweight, but the poison-treated arrowheads make them deadly even for an animal as big as a giraffe.

One anthropologist I met theorized that San trackers purposely hadn't constructed a bigger bow because they feared a larger weapon would lead them to rely less on their tracking skills, changing their relationship with the wild. To me, that spoke of genius, that someone would purposely curtail technology in order to maintain their deep connection with the wild.

I witnessed something similar when I was filming in the Kalahari. One day we watched a group of young hunters preparing to go out into the bush, their dogs leaping excitedly around them, eager to pick up a scent. My San teacher !Nqate turned to me and said, "You see the young people there? You see how they're using the dogs for tracking? Don't let the dog's nose destroy your tracking skills."

Hunting with dogs had become much more common among the San, but some master trackers still refused to use them for that reason.

What I learned from this wisdom is that tools and helpers are well and good, but if you become overly reliant on them, they can divert you from your own intelligence and skill.

Primal Toolmaking

I find great joy in toolmaking—using my hands to craft tools feels like an essential way to connect with a long lineage of trackers. No matter where I live, I make sure to set aside a small space for a workshop. There I keep bits and pieces of wire, steel, bone, and stone for making frames, tools, gadgets, art, and jewelry.

A few years after moving to the Seaforest, I crafted a little multipurpose tool from a piece of heavy metal. Its main use was as an underwater tripod—I drilled a hole through it, then mounted a screw so that the camera could fit on. But I also wore it on my belt as a weight to hold me down when I wanted to shoot close to the seabed.

When I am in my workshop making tools with Tom, I feel connected to our early human ancestors who innovated better ways of doing things using their hands and whatever materials they could find in the natural world. What could be more human than the impulse to invent, to innovate, to make and mend and tinker?

When Tom was young, we made a broadsword from an old car spring we found underwater in the kelp forest. Now he's much better at making things than I am. More meticulous.

Once, he made an amazing knife from an old plowshare, with a beautiful leather sheath. I taught him how to fabricate the metal piece of the tool and guided him to draw the shape he wanted the knife to be. Then he cut the knife out. I helped him grind the blade with an angle grinder, which spit streams of sparks. Then he smoothed the blade with a lot of polishing and sanding. Finally he began work on the wooden handle, carving it to fit the shape of his hand, beating nails into the sides of the handle. For the next few days, he continued to polish the blade until it was ready for a knife sharpener.

While none of our tools are the proper tools of an expert blacksmith, when I use the tripod or the knife, I feel a sense of satisfaction at having made something with my own hands.

Still, I'm careful not to become so dependent on these tools that I ignore the power of my senses.

Connecting with the Threads

Sometimes, while putting the finishing touches on some tool that reminds me of my San teachers, I recall something disturbing I glimpsed in the Kalahari years ago. On a break from filming, I watched as a young priest, just a few years out of seminary and so new to Africa his ears hadn't even been burned by the sun yet, lectured a group of San shamans.

I was amazed that a European with perhaps two or three years of Bible studies under his belt presumed to instruct these African elders who fully understood the complex nature of multidimensional consciousness and had been to the other side of the universe and back countless times. Starting at a very young age, these men would have been introduced to altered states and the nature of reality. They would have developed a deep connection to their gods and their ancestors. Yet these elders sat there with what I interpreted as forbearance as the priest read to them from his Bible.

I've been humbled beyond measure at the grace with which so many Indigenous teachers have shared with me their knowledge of the old ways—knowledge that has the potential to ensure the survival of many species, including our own.

I am particularly honored to have learned from a dear former partner, Charmaine Joseph Gwaza, who took me deep into the spiritual culture of her family and people, and shared with me their rituals and practices. We visited Zululand, where we stayed in the traditional mud-and-thatch huts of her family for several weeks and spent our days collecting medicinal plants.

Life in Zululand let me see what was possible in the world. The people took care of almost all their food, shelter, and tool needs from within the local community—as far as I could see only sugar and oil were imported. Between the bush, livestock, and farming, they enjoyed a good, healthy life, and there was a sense of joy and a slow pace. Laughter came easy.

In sharp contrast, township life a thousand miles away in Cape Town, created under Apartheid, was hard. For a while I lived in the township with Charmaine, and she practiced traditional medicine in a rondavel next to the small house. Several children next door never had enough food to eat, and the noise in the township was earsplitting and constant. It was a heart-breaking and infuriating daily reminder of the legacy of Apartheid and colonialism.

During the three years I lived with Charmaine, we had profound experiences with wild animals. Charmaine was deeply connected to Mdau, the water spirit. The mysterious practice of "calling to the water" occasionally made animals behave in ways I did not understand.

One cold very early morning in Pringle Bay, a coastal village east of Cape Town, I watched as Charmaine sliced her arm with a razor blade and drew some blood. She sat on shore, lit a small pile of imphepho—a type of African sage—and began to pray to the essence of seawater.

She dripped her blood into my right hand and asked me to carry it out into the water. The sea was calm, and I waded out until I was about chest deep. In my left hand I held a live scorpion, which I planned to release back on shore. Charmaine instructed me to pray to the ocean and release the blood as a token of our appreciation.

I did as instructed and began to wade back to land, where I gently released the scorpion. But as soon as I turned my back, I heard a splashing sound.

When I turned around, I was gobsmacked to see a Cape clawless otter floating in the shallow water. Charmaine showed no surprise, as her belief was unshakable and she'd seen many unusual things in her life as a traditional healer. We slowly approached the otter and I saw something I'd never seen before or since: it was as if something unseen was holding the animal in place. The otter was churning the water, snarling, and keeping eye contact with us yet not moving away. Charmaine gestured at me to sit down next to her in the shallow water. As the otter splashed water all over our bodies, Charmaine uttered a series of long sentences directed at Mdau, thanking the water spirit. At that, the otter hurriedly swam away.

My Western education had no explanation for what transpired that day, yet I had a sense that we'd encountered something very affirming and powerful—a feeling that's only deepened after my many inexplicable encounters with this species over the years.

While many of the practices I've been allowed to witness belong to cultures not my own, I try every day to live in a way that honors the spirit of what I've been taught. My daily dive is my most important ritual, connecting me to the ancestors and the place where I come from. Simply giving thanks to our great Mother and asking the ocean to share her wild intelligence whisks me far away from the tame world and into the realm of wonder. Each night before bed, I picture my teachers like Charmaine and !Nqate, who have both passed away, and thank them for sharing their lives and wisdom.

The Power of Ritual

Though Jannes and I have discussed many aspects of Indigenous knowledge as it pertains to our mutual interest in marine biology and tracking, I must admit I was surprised when he asked me to design a ritual for him, a kind of coming-of-age rite of passage—

something every Indigenous culture has but that was missing from the German culture in which he was raised.

A marine biologist with a strong interest in research and ecology, Jannes did his PhD on hermit crabs, and had developed a growing interest in the Cape rock crab. So I designed a ritual that mirrored the experience of a Cape rock crab molting, using methods I'd learned in my years with Charmaine.

Crustaceans outgrow their shells just as humans outgrow their clothing. Since the shell cannot expand, the crab sheds its shell periodically, then grows a new, bigger one. While molting, the animal is extremely vulnerable. It must leave its safe, hard exoskeleton behind, and its body is completely soft until the new under-armor begins to harden. Predators that couldn't touch it when it had a hard shell and menacing claws can now snap it up in an instant. So during the molting process, the crab buries itself deep in the crevices of a cave or beneath the sand, shedding and regenerating its protective layers in secrecy.

It seemed quite natural for us to build the ritual around this crab that kept showing itself to us in many ways. We'd observed its grazing tracks, its ability to swim fast with its paddle-shaped legs, its knack for avoiding octopus predation by running across the kelp heads at low tide, and its miraculous ability to incubate and regrow its legs if lost. Its special claws can hold on to rocks during powerful ocean surges and its grip is four times stronger than that of most other crabs its size.

I dropped Jannes off in a deserted cave next to the ocean, where he overnighted with no food and roamed the shore alone during the day, strangely finding an old shoe whose brand name was the same as his scientific mentor in the US.

The next day, I picked Jannes up and brought him home. Swati, his mother, Anja, and her partner, Harold, and I sang to him as I pierced his skin with a rock crab pincer, then rubbed ground-up crab shell into the cuts. We held him tightly in our arms and he had to slowly push through, mimicking a crab molting, shedding old skin and emerging bright and new.

We concluded the ritual the next day by swimming out to our favorite sea cave, where Jannes swam through the dark cave without a mask and emerged into the light.

Two weeks later, Jannes and I swam out to the same cave as part of our scientific work looking for new species. As we swam into the cave, two Cape rock crabs dropped from the ceiling. A group of scuba divers had just passed through, kicking up a huge plume of silt that made it hard to see the crabs.

When the silt settled and the water cleared, I realized that what I'd thought were two crabs was actually a single crab trying to shrug off its molt. After the fragile, shell-less crab hit the seafloor, it managed to part from its old skin and scuttle away to safety.

I stared at the crab, stunned. In thirty-five years of diving, I had never actually seen a Cape rock crab molting. Yet just weeks after the ritual marking Jannes breaking free of his shell and stepping into a new stage of life, we had witnessed it in nature.

The more time you spend in nature, the more frequently you're thrust into the realm of the mysterious. I've spoken to many nature cinematographers who've shared the kinds of seren-dipitous discoveries that can't all be attributed to attention bias alone. The deeper I connect with wildness, the more it feels like one immense intelligence with which we are somehow interact-ing. We don't quite know what's going on, but it's so much more vibrant and interactive than we think.

Just look at the universe itself: What are the chances of the universe birthing Earth and us and all of these other marvelous life-forms by accident? The odds appear to be very, very unlikely.

To someone raised with a Western scientific mindset, every serendipitous discovery may feel unbelievable. Sometimes our statistical thinking jumps in and tries to calculate the odds. But many of the Indigenous people I've met would never question this. They understand that this is the pattern of life: that nature is alive and intelligent and reciprocal, and that the animals we seek to know sometimes seem to be looking back at us.

The molting rock crab was not the first experience Jannes and I had that defied coincidence. One day, as we were warming up in the sauna after a morning of swimming, I asked him what he thought of this strange phenomenon, of how one often finds questions answered in the strangest ways and in the most unex-pected places.

"It's a bit like falling in love," he said.

I thought this was a good analogy. Love is so unpredictable and mysterious, yet there is a strange and elusive formula to it. It involves being in the right place at the right time, as well as being vulnerable enough to allow a connection to spark.

It's about shedding our protective armor, as terrifying and dangerous as that can be.

Animal and Mineral Ancestors

One morning I walked twenty minutes up a mountain path to a small weir filled with water stained a rich reddish brown by the fynbos vegetation. I was visiting Pippa at her new home at the end of Africa, no farther south to go. Along the path, I encountered a tortoise, a gorgeous big spider and her web, and magnificent sandstone rocks carved by wind and water for millions of years. At the pool, I was greeted by the delicious sound of the amphibious world: frogs.

I envy frogs, the way they can leap from land to air and hold their breath for five hours underwater. I envy the way they can see colors in pitch darkness where I see nothing, not even shapes, and the way they absorb water through the "drinking patch" under their bodies.

As a species, frogs are 350 million years old, while we are only 300,000 years old. They have survived five mass extinctions, and now the sixth human-made extinction is seriously threatening many frog species.

Submerged in water up to my neck, I slowly walked along the oozing bottom, enjoying the feeling of it on my feet, careful not to damage my toes on the rocks. A large dragonfly larva swam past, and I admired the first flush of new wings on her back, the impressive mouthparts that crush her prey. She slid into my hand, this creature from three realms: water, then land, then air.

She bit my skin, a sharp pain but no blood, and jetted off into the dark water.

I followed the sounds of the frogs and was delighted to find a hollow deep inside the bush bank. There I counted eight Cape river frogs, large eyes staring through me. I tried to put out a nonthreatening feeling and gently eased closer. I felt grateful that they allowed me so near their haven. I'm not sure why they didn't jump into the safety of the water, but they sat still, their spotted throats moving, eyes tiny galaxies of gold, red, and black, massive leg muscles leap-ready.

And in that instant, suspended in the embrace of the water, staring into eight pairs of unblinking eyes, I had a simple yet profound insight: the tortoise, the spider, the frogs were not only my kin; they were my *ancestors*.

I had literally been born out of their millennia-old lifespring. What's more: we all are born out of the substance in the giant rocks on the path I'd just walked up. The same minerals in the rocks are inside our bodies.

This was not some fanciful idea I'd dreamed up; it was inspired by the work of Brian Swimme, a brilliant author and professor of evolutionary cosmology at the California Institute of Integral Studies.

Meeting Brian through my old friend Professor Louis Herman was deeply inspiring. It was like meeting a fellow adventurer who had been diving his whole life, but rather than immersing himself in a small stretch of Seaforest, Brian had been swimming in our entire universe. I was amazed to discover that many of the teachings I had received from animals and nature were eerily similar to the insights Brian had gleaned from his deep study of galaxies and exploding stars.

In his book *Cosmogenesis*, Brian shared the epiphany that our minds are the creation of all the minds that came before us. "The ideas of Galileo and Newton and Einstein shape my daily perceptions," he wrote. "If you take this further, it means our ancestors constructed all the human minds on the planet today. Even though I regard my mind as my own, the fact is, others have built it."[4]

So in a way, we are living downloads of our ancestors, both animal and human.

The ideas of the African geniuses who invented symbols and left them in Blombos Cave so many thousands of years ago shape my daily thoughts and feelings. I can almost feel that ancient hand covered in red ochre guiding my fingers as I write this. The primal parts of my mind—things like sensing danger—were shaped by my animal ancestors. Other parts were formed by ancient hominids. When I cook my food over a fire, as I do every week, that part of my mind is drawing on my *Homo erectus* ancestors from half a million years ago. When I make more sophisticated tools, my hands are being guided by my *Homo sapiens* ancestors from 100,000 years ago.

As Brian says so eloquently, the whole universe is inside of us and we are inside of it.

Crouching down in that little frog hollow, I felt deeply connected to all living creatures, animals, plants, and the earth and elements that sustain us. We are all made of the same substance, and that's why in moments of grace I feel so woven in. I feel frog, I feel dragonfly, and if I stretch my mind, I even feel rock.

I'm a frog-rock-spider-dragonfly. I'm a galaxy, and so are you.

The Octopus Village

After many months with few octopus sightings, there came a time when I started to see them everywhere. It was still a mystery how and why they all returned to the shallows in early summer. There is some suggestion that as they age, octopuses move offshore and return for mating.

Jannes thinks that they don't like turbulent water, as it fills their dens with sand and breaks down their den walls. So maybe in the rougher months during winter, when we have more storm surges, they move into deeper water where it's calmer, and return to the shallows in summer when the storms subside. I've noticed they prefer sheltered bays where their dens require less maintenance, which in some ways supports this deep-water idea.

Swimming on the surface, I searched for octopus kills. I was by then swimming more than a mile in a stretch, moving through the kelp forest with strong, easy strokes. I had developed a swimming technique to create minimal pressure waves in the water so as not to disturb animals, carefully dipping my hands in without splashing, and not kicking at all.

I found signs of octopus predation everywhere. It's as if they were teaching me about all the other animals who inhabit each section of the Seaforest: chitons and abalone in the thick kelp shallows; swimming crabs in the open sandy areas to the west; otter shells in the eastern sandy areas; brown bivalves in the deeper edges of sand and kelp; and helmet snails in specific rocky areas full of urchins with surrounding sandy nodes.

It's extraordinary how versatile the octopus is. Each individual targets the most abundant prey in their immediate area and figures out the best ways to catch this prey, thereby becoming a specialist. As I swam from den to den, I created a map in my mind I called the Octopus Village, a map that showed me where many hidden animals were abundant.

Jannes and I spotted a subadult pyjama shark scenting something under a rock. When we looked closer we saw that its target was a small octopus. Subadult pyjama sharks are especially dangerous to young octopuses because these slim sharks can push their noses and teeth into the tiniest cracks.

Then we saw something surprising. The shark almost grabbed the octopus, but the clever youngster was holding a piece of slippery orbit—a tough, olive-brown algae. The octopus pushed the algae into the shark's mouth and the shark gulped it down, giving the octopus precious moments to dig under the sand and escape. We watched as the outwitted "vegetarian" shark glided away, cheated of its octopus meal.

I remembered suddenly that the week before I'd seen an octopus chewing on a piece of kelp, its frond trailing out of the den. My mind octopus-curious, I'd looked closer. She released the kelp, and as it floated away I noticed bite marks matching the shape and size of her beak. To my knowledge, octopuses had never been observed eating algae. I wondered at the time if the octopus was just bored, or if the kelp had some nutritional or medicinal value.

Now I reconsidered. Was there something else going on here? Was it possible that the octopus was using the kelp as a *tool*?

These pieces of chewed kelp had the same rubbery texture and scent as octopus flesh. They could be used as tools to fend off and confuse shortsighted predators like sharks, who rely on smell and taste and texture.

It's quite remarkable to think of a mollusk using a tool, especially if that tool is kelp. Octopuses certainly use kelp to wrap around their vulnerable bodies to create a type of armor, but if my hunch was correct, what I was seeing was even more extraordinary—a sophisticated subterfuge!

This creature has no limits when it comes to teaching me and expanding my mind.

I eagerly shared my findings with Jane Goodall, whose breakthroughs on tool use among chimpanzees transformed our understanding of animal intelligence and behavior. Whereas once the ability to make and use tools was thought to be the domain of human beings alone, in 1960, Jane observed a chimpanzee she named David Greybeard using straw to scoop termites out of a mound and into his mouth.[5]

Her finding was followed by a wave of other discoveries about the deep ingenuity of our animal kin, from dolphins using sponges to protect their beaks while foraging to crows fashioning hooks out of twigs to elephants' methods for making drinking holes (they dig the holes with their tusks, use chewed-up balls of tree bark to plug them, then cover the holes with sand to prevent evaporation so they can return to the spot later).

I can't help but wonder if our discoveries about tool use in octopuses could transform our views on this animal's intelligence and abilities, just as the finds in Blombos Cave changed our understanding of the cognitive abilities of Middle Stone Age *Homo sapiens.*

An Amphibious Tree

At the entrance to a cave perched high on a cliff overlooking a raging sea, I could see the wind rippling across the great bay, lifting tongues of sea spray high into the air. Dark clouds and distant rains served as the background for patches of silver ocean where the sun was trying to penetrate.

The cave floor was covered with shells, the last meals of the wild people who once lived on this shore. Careful not to crush any shells, I climbed under a shelf and belly-crawled deeper into the cave. The gusts of wind rushing around me gave me a strange sensation, activating my sense of pressure: something I often feel underwater, but rarely on land.

I crawled out of the small space and walked over to three small wild camphor trees. Unlike other camphor trees I'd seen, these had the strangest trunks, thick and bulbous, with smaller branches coming off the bases, giving them a look similar to bonsais.

Did the last wild people who lived in this cave break off these branches? I moved my hands through the mass of branches, and deep inside me I felt those old hands covered in thick calluses. Was this just my imagination, or was I tapping into ancient memories of humanity's deep connection to these trees?

The camphor tree was certainly important to early people as both its leaves and bark had medicinal properties.[6]

This tree had so many uses for wild foragers that I would not be surprised if it was elevated to a kind of spiritual status. It was particularly significant to me that the trees were growing here at the cave entrance, because the flowering fruit tufts were used by hunter-gatherers as tinder for their fires.

Still, could these three trees be old enough to have been sculpted by ancient hands? And then a memory hit me. My friend Tony Cunningham, a brilliant ethnobotanist, once told me about trees that can survive thousands of years by a process called "clonal resprouting."[7]

Instead of reproducing through seed dispersal, they send out limbs to reproduce and can survive for millennia—one lime tree in northern Scotland has survived an astonishing four thousand years through this method.

Could I be looking at the tracks of ancient hands in the gnarled and twisted shapes of these trees? Or was their unique growth pattern simply evidence of modern fisherfolk gathering firewood?

When I asked Tony, he confirmed that wild camphor is indeed a clonal resprouter and these could be very, very old. Inspired, I decided to look further into this particular species of camphor, *Tarchonanthus littoralis*.

In some ways it's an amphibious tree—adapted for journey by sea. The fruity part is waxy and fluffy and carries the seed. These seeds are blown along the coast and only survive because they don't sink. The waxy waterproof fluff is like a little boat that travels miles along the shore and is deposited high up in the fertile kelp line, where the waterborne seed grows into another camphor tree, resistant to the winds and dryness of this place.

The wild camphor tree had grown a type of mythic status in my mind, and each time I visited the cave I looked forward to greeting these hardy trees, and remembering our deep ancestors who lived so lightly on this shore, who knew by countless generations of trial and error the ways of trees and animals.

I had even done a few experiments with camphor tinder with Tony, but something was always missing. The camphor tinder is a magnificent holder of a glowing coal, yet without an accelerant, it will not break into flame. Then I found the washed-up skull of a Gray's beaked whale while scavenging on the beach. Perhaps whale fat was the missing ingredient? I decided to put my fanciful theory to the test. I managed to collect a small vial of whale wax from the skull and mixed it with camphor tinder. I laid the mixture in an abalone shell and then held my breath as I struck the flint stone. As if in slow motion, the sparks hit the downy tinder drenched in whale wax and instantly ignited. Watching the tinder burn with a strong flame for several minutes, I felt very close to my deep ancestors.

My passion is remembering some of their genius, and through tracking bringing some of their stories and practices back to the surface. I dream of the white smoke in those ancient throats, their laughter, fire flickering eyes, and the amphibious seeds floating on the meniscus of human consciousness, just waiting to wash up on fertile soil, and keep resprouting into the deep future.

A Story Emerges

The more wondrous my encounters with ancestral tools and traditions, and the more I deeply connect with the animals, plants, and minerals that were here for millions of years before me, the more I view tame-world technology as a curse—distractions that pull us away from nature and from the ways of our ancestors.

But I'm also aware that these tools have transformed my tracking practice. The camera gives me a photographic memory, and my computer helps me analyze what I see. I am fast-tracking my learning by creating videos that I can watch again and again, uncovering new clues to mysteries I've been pondering for years.

Modern technology also makes it easy to connect with a team of experts to amplify our collective knowledge. I can bring each new biological mystery to Jannes and Charles or run new theories about early *Homo sapiens* by Christopher and Karen, and have an answer in days, sometimes hours.

I often lose myself in this line of thinking upon waking at 4 a.m., when I think best. Other mornings I delve into fascinating research or pore over the footage of my encounters, awestruck at what the natural world is teaching me.

For years, I used my camera not as a filmmaker but rather as a tracker. It was simply a highly sophisticated tool for helping me better understand the ecosystem. The footage I was amassing was awe-inspiring, but after years of immersing my body and soul in the healing waters of the Seaforest, the last thing I wanted to do was wake the muse again with another film project.

But all that changed when, after four or five years of miraculous encounters with the creatures of the Seaforest, I began to see signs of an incredible story emerging.

CHAPTER SIX

FEAR

SOMETHING ROUND CATCHES MY EYE. I LEAVE MY FATHER'S SIDE and dive deeper to get a closer look.

The size of a dinner plate, the creature glides slowly across the sandy seafloor, the edges of its fin fluttering delicately as it moves.

I want so badly to touch it.

I reach out . . .

My regret is instantaneous as electricity floods my body.

MORE THAN FORTY YEARS LATER, MY FIRST EXPERIENCE WITH THE onefin electric ray sits strongly in my memory. Electric rays can produce 220 volts of electricity—and the shock I'd felt that day was a lot like the experience of sticking a finger in an electric socket. Years would pass before I'd be able to get close to that animal again.

But it didn't stop me from feeling the pull toward other animals known to be dangerous to humans. For as long as I can remember, I've reached toward nature with little thought to the risks. For better or worse, my desire to be close to wildness has always beaten out my fear.

For years I fed that desire not only with daily dives in the Seaforest but also with regular visits to the Jukani Wildlife Sanctuary, formerly located about four hours from Cape Town. At the time of my visits, the sanctuary specialized in rescuing and caring for animals that were born in captivity and would not be able to survive in the wild.

Visitors were not allowed to interact with the cats, but after years of filming their work, the sanctuary's former owners, Jurg and Karen Olsen, gave me the rare privilege to film in areas where certain animals lived in large enclosures.

Of all the big cats at the sanctuary, none enthralled me more than the jaguar. One afternoon I was led into the sprawling grassy habitat where a female jaguar roamed. What most stood out to me about her were her powerful muscular body and huge golden eyes.

I had not been in her presence long when a danger presented itself: a portion of a tree branch had caught and pinched her tail. Cats transfer aggression, and I knew her pain could easily lead her to lash out at me.

As she hissed and snarled in annoyance, I froze. I was mesmerized by the size and thickness of her canines, which were cone-shaped and razor-sharp. Jaguars have the strongest jaws of any big cat—their bite force is about double that of a tiger's. Unlike other cats, who grab their prey by the neck or spine, jaguars kill by biting right through the skull of their quarry.

I was terrified.

But I'd been in similarly frightening situations before. My mind worked quickly. I knew from my time with wild predators like crocodiles and black mambas that movement would only upset her more, so I kept dead still.

This was not easy.

I was just inches away from her, and the sounds rumbling deep in her throat seemed to vibrate throughout my entire body. I prayed that I would be okay, but within my prayer was a kind of acceptance. I knew I had no control over her actions and could only surrender to the moment.

And then something astonishing happened. Her tail came free, her eyes softened, and the electric tension in her body released. She stopped hissing, and the pause gave me a chance to make myself smaller and bring myself down to her level. As I did, she lifted her big paws over my shoulders, her touch surprisingly gentle. She was ten times my strength and could crush my skull in a second, yet her paws rested lightly on my shoulders.

As a rule, I never touch an animal unless they make first contact, but I instinctively returned the embrace. I could feel my body shaking slightly from the shock. I felt many things in that moment: awe, gratitude, wonder.

It was not unlike the first time I dived with great white sharks. Five of the animals made famous by the movie *Jaws* had circled around me, but up close they weren't monstrous at all, just powerful and beautiful. Each animal weighed over two thousand pounds, and they had large dark-blue eyes and massive dorsal fins. They moved slowly, gliding, seemingly propelled by an unseen force.

After some time one of the five white sharks opened its mouth in an action called "gaping." This told me that it'd had enough of my presence, a first gentle warning. I slowly backed away and exited the water.

Back on land, my whole awareness of fear had dwindled so much it was difficult even to put my seat belt on in the car. Every time you tackle some new primal fear, your other fears subside.

That can be transformative. It can also be perilous.

The Life Givers

For me, managing fear—any fear, but especially of wildness—is all about getting to know whatever it is you're afraid of. It's also about getting to know your limits. Most fear is of the unknown, after all.

Get to the root of some of our species' most common fears and you'll find intelligence there—a brilliant survival mechanism that helped wild people and animals survive. Consider a fear of deep water—the sea can be very dangerous until one gets to know her moods, so there is wisdom in approaching her with caution.

Once you understand her ebbs and flows, however, you can make decisions that factor in the force of the wind and tides, the temperature of the air and water, as well as your own strength and energy levels. What was once foreboding becomes a powerful friend that you treat with respect and care.

Fear of diving in a place where sharks swim is very tangible and common, and one I've of course experienced, but what's helped me overcome it is a rational assessment of the risk involved. Animals are seldom aggressive unless provoked or cornered. In fact, it's much more likely you'll be injured by a toaster, a chair, or a coconut falling from a tree. Even driving to the sea in a car is thousands of times more dangerous than sharks.

Yet the primal fear of these predators has a powerful grip on the mind.

Our fears connect us to our wild animal kin, who experience fear and danger all the time. Primal fears are life givers, even though the things we fear occasionally take a life. The great irony here is that the forces of the wild—the cold, the dark, big predators—are the very things we need to invigorate our beings.

Whether danger is real or just *feels* real, fear is a part of being alive. And a crucial part of being wild is not running from experiences that are scary or challenging but surrendering to this natural biological process, to the parts of life we can't control, learning to sit with fear, breathe through it, honor it.

The tame world wants to control everything, but nature is not controllable.

Nature *is* something we can get to know better—not with

the aim of controlling it but with the goal of training ourselves to approach it with as much clarity and calm as we can muster. That's why the cold can be such a powerful tool. It is frightening and dangerous, but relatively easy to deal with in a safe way. You can get used to the cold with ice baths or short dips in the ocean followed by warm clothes and hot baths.

Your central nervous system doesn't know the difference between controlled and uncontrolled cold. It still responds the way it would in a life-threatening situation—so when you make it through to the other side, your entire relationship with fear seems to have been altered.

Of course, if you panic or get very stressed or you're breathing in heavily through the mouth, the cortisol kicks in. Triggering that primary stress hormone can lead to sleeplessness and anxiety. But with time and patience you can, breath by breath, familiarize yourself with primal fears like cold, darkness, and deep water so that your body recognizes the sensation and responds with calm.

I have never thought of myself as fearless. For all the dangerous situations I've found myself in over the years, I still carry within me that little boy who was terrified of the things he heard and saw in the darkness. I've brought him along with me on great adventures, and thought about him when my son, Tom, faced his own fears of the dark as a child.

That little boy within watched me dive deeper and deeper into the dark, like a dreamer watching a dream.

Until one restless night, he woke up.

The Seeds of Inspiration

When I began the journey to reconnect with my amphibious soul, I was sure of one thing: I never wanted to make another film again. Completing three crocodile films back-to-back, the decades of travel and editing, and the inevitable crash after each film had taken their toll. The years seemed to blur together into a picture of exhaustion and burnout.

That's it, I thought. *I've had enough*.

All I wanted to do was immerse myself in nature.

In those first few years after returning to the Seaforest, I found other ways to celebrate the wisdom and wildness of Africa. I teamed up with a small group of like-minded individuals to launch an organization called the Sea Change Project, which is dedicated to conserving and protecting the Great African Seaforest. Inspired by Indigenous wisdom, we combine science and storytelling, and favor lived experiences with nature as our teacher.

After all these incredible experiences, it began to dawn on me that I was not the same person who'd washed up on this coast just a few years earlier, body and spirit broken. Cold water tracking had restored my health and sharpened my mind. My relationships with the animals of the Seaforest had given me a sense of hope and renewed energy, even on the days when I felt despair about looming environmental crises.

As time passed, my feelings about filmmaking, too, started to shift. The more I pored over my footage, the more I started to feel a familiar feeling: the seed of inspiration taking root. I wanted to share the stories of the Seaforest with others, to take people inside this marvelous ecosystem I'd discovered. In particular, I wanted to share the growing friendship I was cultivating with a curious octopus who showed me so much about her world.

One afternoon I showed some clips to my friend Roger Horrocks, the gifted cinematographer who had helped me bring the crocodile films to life. He agreed: here were the makings of a remarkable story.

The desire to introduce the creatures of the Seaforest to the wider world felt like a natural progression of my exploration of wildness. For eons, our ancestors had gone out in the wild, then come back to the village with stories to share. They had recorded adventures of the body and spirit using ochre on cave walls. Those images have endured and continue to be seen by generations of humans. In my own way, I felt part of this long continuum of artists and storytellers. After believing I'd need to leave my art behind in order to feel balanced and stable, this captivating story had emerged, and I felt compelled to share it.

I also suspected the process could look very different from that of my previous films. I wouldn't have to leave my home and family behind to spend weeks or months at a time on location. I already had thousands of hours of footage and a brilliant team. Though she lacked filmmaking experience, I could see in Pippa both a passion for wild creatures and a knack for storytelling.

It was Pippa who had the idea not to make this an activist's conservation film but simply to tell the true story of a wild friendship. Roger expressed an interest in helping to film, using his sophisticated camera systems. And Swati was always there behind the scenes, keeping everyone on track with her wisdom, kindness, and unwavering strength.

If I had any fears about embarking on this journey, about the terrible feeling returning, they were very far away—under the surface, out of reach.

Meanwhile, my friends and I continued to face our primal fears.

A Glimpse of Magic

When I first began meeting the creatures of the Seaforest, I wanted so much for Swati to join me on my morning dives. But for the first years of our relationship, the prospect of even a short swim together was unthinkable.

When she was a child, a swimming teacher had pushed her to keep her head under water, making Swati feel she was almost drowning. As a result, she never learned to swim, and for a long time, even just putting her head under the water triggered that deep-rooted panic.

Years later, she would have a profoundly terrifying encounter with the ocean—twelve years before we met, she was in Sri Lanka when an earthquake off the coast of Indonesia triggered the tsunami that killed over 230,000 people, including 35,000 in Sri Lanka. She awoke to the sound of a roar like a freight train in a tunnel. She looked out the window of her hotel and saw the ocean was higher than the trees. Her room filled up with seawater that reached all the way to her neck before miraculously subsiding. Most other hotels in the immediate area were washed away, leaving few survivors.

I understood the seriousness of her fears, of course, while also wanting to do what I could to help her feel more at ease. I offered to share what I'd learned—how to slow my breathing, how I could rely completely on the buoyancy of my full lungs. I encouraged her to come out on calm days and assured her I would stay with her to keep her safe. But she wasn't having it.

"I know you sympathize with me, Craig," she said, "but you don't understand because you've never felt this kind of fear. You learned to swim before it was a conscious thought."

She was right. I didn't know what it was to fear the ocean, because I felt nearly as at home in the sea as I did on land. But Swati hadn't been placed into the ocean as a newborn; this place had not always been her home as it was mine. I hoped she would grow to love the Seaforest, but I couldn't force or control it. Her relationship with the ocean was her own. I had to trust that it would grow and deepen in its own way and in its own time.

On the Other Side of Danger

While my relationship with the sea is a longstanding one, it has had ebbs and flows. When I was young, I did not always respect my limits and I lacked the understanding of the ocean that I have today.

One of my first harrowing experiences with the ocean was three decades ago. My brother, Damon, and I and my best friend from high school, John, were diving in the kelp forest when an enormous set of waves came in.

As the first wave barreled toward us, I dove down about fifteen feet and held on to two thick kelp stipes. The mighty force of the wave ripped the kelp off the seafloor, sending me soaring into the reef. I heard a strange sound and realized that the waves were so powerful even the giant boulders on the seafloor were being tossed around.

Eventually we managed to swim far enough out beyond the waves. We knew our best option would be to swim parallel to the coast and exit the water at a beach or flat rocky area. But we encountered a very strong current. For well over an hour, we struggled to swim against the current, because we knew if we went with it, we'd be smashed to pieces against the high sea cliffs.

We were young and fit, so we made a little headway, but the body has its limits, and after another half hour, we all began to experience severe leg cramps.

We stopped talking and just swam, one stroke after another. We didn't have the energy for words, only our grim determination to stay alive. I was very frightened, and I felt sure that not all of us would survive. The cramps became so painful we could no longer keep swimming straight, and we realized we needed to make for shore before we drowned. It was a horrible feeling to be so exhausted and to know that at any moment we could be flung toward the rocks.

Then I looked back and saw a monstrous wave, close to twenty feet in size—hundreds of tons of water moving at great speed. I can still recall the fear I felt when I saw that giant wall of water coming toward me.

As the wave hit, it picked us up like twigs. I prayed to everyone and everything I loved and curled up in a ball, protecting my head with my arms. Then, as I held my body tightly, the wave lifted us over the cliff, took us up the mountain slope, and deposited us in an acacia tree.

The frothing mass of water receded, leaving the three of us stranded in the branches about twelve feet above the ground, exhausted and drenched, but miraculously unharmed. Too dazed to speak, we climbed down out of the tree, tested our weak legs on the ground, and tottered back toward my battered old car and into life. I felt very mortal, and put my seat belt on.

For months afterward, I entered the water with some fear, cautiously and respectfully. I avoided rough seas and areas with strong currents. I used the thick kelp to protect me. I learned to read the weather and to notice changes in the atmosphere that might affect the ocean. And I constantly scanned the horizon for waves.

I'd felt the raw power of this ocean and I was determined to work with it, not against it. The ocean had been totally masterful in its authority, my body like a tiny cork in a raging torrent, with no control, completely at its mercy.

Primal Joy

It was during a trip to Pringle Bay, where my parents live, that Swati decided it was finally time to confront her fear. We'd just come back from a couple of weeks in the Kalahari. It was winter, and the water was barely above 50 degrees Fahrenheit, so she wore a wetsuit, gloves, and hood, but she still found the cold excruciating. She bravely waded into the shallows—and then she turned around and came right back out.

"I've never felt this kind of pain in my life," she said, gasping for breath. "It's like being stabbed with knives!"

She tried again on a warmer day, determined to push through the discomfort and the fear. Using a snorkel to help her breathe and a life jacket, wetsuit, and fins to keep her afloat, she followed me out in the water. I tried to stay close enough to help her float, but not so close as to stifle her sense of discovery. Making sure her mask was well sealed to her face, she dipped her head beneath the surface of the water for the first time since she was a child.

Scientists are beginning to understand how, when we have a traumatic experience like Swati did, the body remembers. I suspect at that moment her primal intelligence was screaming at her that she'd been here before, this was dangerous, and she needed to escape.

But that's not the only message Swati was receiving that day.

The snorkel and mask allowed her to see an undersea world that was entirely new to her, full of animal kin. Gliding beneath us through the kelp were several spotted gully sharks, beautiful sharks about five feet long that posed no threat to us. Swati turned her head to watch them swim by. For several long moments, we watched the sharks, mesmerized.

Back on shore, she shivered uncontrollably as I helped her warm up.

"That was incredible!" she said, teeth chattering. The sharks had given her a glimpse of the magic I was always telling her about.

"One thing's for sure," she said. "I will never be able to do this without a wetsuit."

AFTER THAT DAY, SWATI ENTERED THE WATER MANY MORE TIMES, learning to float on her own with a mask, snorkel, and fins. At first, she would go in only as deep as she could stand or where she could hold on to the kelp. In time, as she grew more confident in the water, she began to have her own special experiences that inspired her to keep going. As long as there were animals to see, she felt invigorated and wanted to stay longer in the water. Without the presence of wildlife to soothe her soul, the old fear would start welling up again.

One day we encountered a group of short-tailed stingrays, the largest stingray in the world, which grow to some fourteen feet in length.

While their venomous tail barbs can be lethal, the rays are not aggressive and only use their barbs for self-defense. Seven of them were gracefully soaring around us, the first time Swati had ever witnessed such a sight. We stayed in the water watching the rays all morning. She later told me she was so enthralled she'd lost all fear of drowning. Little by little, primal joy was taking the place of her early trauma.

Facing the Darkness

Pippa, Tom, and I were walking along one of the oldest human pathways on Earth, a place where our earliest ancestors once walked. We still find artifacts in their caves stretching back to the Stone Age. The narrow, rocky path, which is colored red from ochre, spanned a huge cliff overlooking False Bay, and it felt like we were walking in midair.

This great bay is the Serengeti of the Sea, traversed by whales, orcas, dolphins, and huge fish shoals. By some miracle of fecundity, the bay is still full of life despite three hundred years of hard fishing.

I'd trod this ancient path many times, but this day was something special, a perfect storm of events that allowed access to a very rare place, and I was excited to share it with my son and our friend.

We descended giant vertical sandstone steps into the crystal-blue ocean. Each step was so tall and steep that we had to clamber down very carefully, as a fall would be fatal. We passed our packs down, knowing it would be easier to climb without the extra weight, and all felt relief as we reached the edge of the sea.

The cliff continues underwater into deep water and a thriving kelp forest filled with invertebrates. We walked around a small headland, and there it was: the sea cave of secrets.

The cave announced itself with a thunderous *boom!* caused by rolling waves and the release of air trapped in the dark interior. The cave was huge and foreboding, and I had a scary memory of a visit when the waters were not so calm and I was flung about in the back before managing to escape, scraped and bleeding but not broken.

Normally the water is extremely murky—sediments collect and are forced into the water column by the fierce currents. But a rare upwelling of 53-degree water had flooded the cave the day Pippa, Tom, and I arrived. I could see fish rising from the depths and grabbing food from the surface.

"It's so clear I can see that rock crab on the bottom," Pippa cried.

I had never seen the water like this, and I knew we might never get another chance this good.

"Guys, this is really special," I told them. "I think we should try to go deep into the sea cave."

Tom was quiet—I sensed he was tense yet curious to enter the inky blackness.

We put on minimal gear. For me, wet shorts with pockets for a small camera, lenses, and my high-powered underwater torch, plus a mask, hoodie, and snorkel. No fins or wetsuits. We slipped into the water, trying not to create pressure waves that would alert animals in the depths, then we swam toward the deep darkness of the cave mouth.

It's such a powerful place, here on the edge of Africa. The cave itself felt alive, a creature of unknowable age excavated over millennia by the battering of millions of swells that carved deep into the towering cliff. The pungent stench of cormorant guano mixed with the slightly fishy scent of otter spraint.

We were immersed in primal sensations: the icy water, the wild smells, and the vision of swimming into the giant cave creature's mouth. The light reflected off the seafloor, and dancing on the surface were hundreds of cigar comb jellyfish, their cilia creating iridescent light shows.

We were entering a fantasy world that was somehow real. My flashlight looked like a light saber penetrating the darkness and illuminating the giant mollusks, anemones, and urchins living in the dark. I found a huge underwater passage big enough to park several cars. We followed this tunnel deep into the cliff and looked back at the entrance far away. The vaulted ceiling shimmered with reflections from the water and my flashlight, rippling snakes made of light.

We turned around and swam to the deep place where I'd been thrown about before. The water was so black I could hardly see my hand in front of me.

My primal mind saw this darkness and screamed, *Get out!* In past times, especially in rivers, big predators would have used their superior sensing systems to "see" us in the dark water and pick us off like ripe berries.

I registered the fear and thanked my primal mind for the warning. You can befriend the primal mind, speak to it the way you might your beloved dog who barks at a visitor. *Thank you, but there is nothing to worry about today.*

It was so calm and clear, I couldn't believe our luck. In the pitch dark my flashlight beam picked up a large dark body moving gracefully through the water.

A predator, my primal mind warned.

Yes, I thought, *but also a friend.*

It was an adult gully shark, followed by a younger one. Tom and I have dived with these sharks since he was tiny—they are like old trusted friends. I noticed fish too; I suspect the sharks may have been hunting the fish by trapping them in the back of the cave.

As we swam from the darkness of the cave back into the light of day, the dark water in front of us turned silvery blue, the surface shimmering with golden sunlight. The cave had spoken with us, whispering its secrets, and we emerged changed, made fresh, washed clean in nature's cauldron.

We warmed our bodies by lying in the sun on the rock shelf at the edge of the cave.

"Listen to those water drums," Tom said. He pointed at the cliff and gestured. "It's from little wavelets catching air under the rock ledges."

We boiled mussels and feasted on their freshness. Then we climbed up the cliff and walked back on one of the oldest paths on Earth.

We were high on the grace of nature, deeply thankful that this perfect window had opened up for us and allowed us to see inside the creature that is the sea cave of secrets.

Fathers and Sons

Several weeks later, Tom and I were walking toward the water close to home.

"Since we went to the cave, I'm not afraid of the dark anymore," he announced.

I was a little surprised to hear this. I knew he'd been scared of the dark when he was younger, and we'd often talked about it. I'd told him that I'd experienced the same thing when I was his age, that I understood.

But Tom never had said he was still wrestling with the fear at age seventeen. My mind flicked back to the time in the sea cave, the tension in his body when we approached the dark entrance.

I felt a flood of conflicting emotions. Disappointment in myself at not having realized my son was dealing with something so challenging. Proud of him for diving into the darkness. Even more proud of him for sharing his fear with me. My son was growing into such a wonderful person.

I wanted Tom to feel like he could talk to me about any difficulties he was facing. I thought back to my own childhood and how much I had kept to myself when it came to my fears about darkness, my shyness around people.

My father was so strong, he had seemed immortal. I had wanted to make him proud and had been eager to follow him into the crashing waves, even when they rose high above my head.

I wanted to make Tom proud too. To show up for him whenever he needed it.

But it was getting more challenging as the pressures of the film ramped up. Some nights I barely slept at all, and my strength was not what it once was.

Though I knew that our trip to the sea cave of secrets was the kind of experience that made life worth living, it was becoming harder to find the time and the energy for such adventures.

Inferno

Through our back window, I looked out at a strange orange sky. It took me a moment to realize what was happening: there was a big fire upwind on our mountain that rose from the sea. Soon the wind began to carry the smoke to our neighborhood, and eventually the flames too.

As the fire began racing down the mountain, it was terrifying. As soon as it reached our neighborhood, I heard explosions that I surmised were pipes exploding. All around me, parts of houses had begun to melt.

In what seemed like mere moments, our neighborhood was ablaze. Monica's house next to ours was a raging inferno. Fireballs were flying through the air. Pine trees were exploding. The home of my neighbors Mark and Liz would literally vaporize, leaving nothing but a smoldering foundation.

Several helicopters were flying overhead, scooping water from the ocean in giant buckets, then dumping it over the flames, but still the fire kept coming. It became clear we should evacuate. Swati gathered our three cats, our passports, and our underwater footage, then drove to the shore, clearing the roads for the fire trucks approaching. Thankfully, Tom was at his mother's that day.

I felt slightly out of body, hyperalert, as I wrapped cloth around my mouth to reduce smoke inhalation. As I looked at the flames that were consuming our neighborhood, I had a strange thought:

This sounds and feels like a dragon. Visibility was so poor—it was like being underwater—that the sounds of the growling flames, cracking wood, and exploding pipes built the creature's form in my mind. I realized the fire was so big that it might take some time before the fire trucks could come to our aid.

I felt a strange kind of calmness take hold, then I made the decision to stay back with my neighbor and frequent diving partner Andre, a tough and cheerful outdoorsman with whom I once paddled the great Gariep River. He and I broke down the door of Monica's blazing house next door to see if there was anything to be done, but we immediately realized it was impossible.

We were completely surrounded by fire, the heat from the inferno next door overwhelming. Though our home was a vulnerable wooden building, its saving grace was our rainwater tanks. All the municipal pipes had burst in the fire, but our tanks survived, which allowed me to douse the flames and small fireballs that threatened to ignite our place too.

It felt like a bit of a miracle that our house was spared and no one was hurt.

When Swati returned home later that day, we spoke about the events and comforted the shaken cats. Swati told me they had been unusually quiet and still in the car, as if they knew it was life or death.

I think we were both in a state of disbelief. I carried the bag of film back to my office, then Swati made us some tea—we would stay up all night to guard against flare-ups.

As we watched the embers spark and catch fire repeatedly, I

was reminded yet again of how little control we have over wild nature.

The Really Terrible Feeling

It took us almost three years to edit the octopus footage. I had amassed extraordinary shots in the years before we knew what we were filming, and as the story began to take shape, Pippa, Roger, and a talented cinematographer named Warren Smart stepped in. My old friend and collaborator Ellen Windemuth joined as executive producer and brought aboard James Reed, a dynamic British filmmaker. Rounding out the team were brilliant editing consultant Jinx Godfrey; Sara Edelson from Netflix, truly a force of nature for the project; and Swati, who helped guide me in her special way every day.

So much joy went into the filmmaking and editing process, even if hours of extraordinary footage never made it into the final cut. Making this film was such a gratifying experience. Although we were spending hours and hours editing, we stuck to the protocol of diving and tracking every day, preserving our energy and allowing us to work for so long. I did eventually get exhausted but I managed to not entirely succumb to the terrible feeling that had stalked me during my previous work by stepping back and letting the world-class team finish the film.

My Octopus Teacher succeeded far beyond our wildest imaginings. After decades of making documentaries about the people and animals of Africa that rarely reached a big audience, suddenly our little film was being released in nearly 200 countries. Plaudits were pouring in, media outlets were clamoring for interviews, and we started to hear rumors of awards and prizes.

Just as I was experiencing the greatest professional success of my career, the Really Terrible Feeling set in.

Against My Nature

Everyone has their biggest fear. Yours might be fire or falling from a great height, Swati's might be drowning, and I would soon discover my own primal fear. After decades of filmmaking, I'd stepped on the other side of the lens. And now millions of people were looking at me.

I was under a spotlight, being examined from every angle, like a sea creature in a petri dish. I had no armor, no camouflage. Unlike my octopus teacher, I could not cover myself with shells, change color or shape, or cloak myself in a piece of kelp.

I've faced giant walls of water coming at me, felt the bone-tingling hiss of a big cat ready to pounce, broken down the door of a burning house, and dived into the lair of one of the planet's most dangerous apex predators, but any fear I felt in those moments paled in comparison to what I experienced in the weeks and months following the release of our film, when I felt the gaze of millions of people fastened upon me.

FOR MUCH OF OUR EXISTENCE, MOST HUMAN BEINGS WOULD HAVE lived in groups of about thirty other people. Of course, life is quite different today, when social media connects us to hundreds, thousands, millions of people we'll never really know. But the ability to connect with, and be seen by, tens of thousands of strangers has only existed for an eyeblink in human history.

For most of our deeper evolution we existed in societies similar to the San hunter-gatherers of recent times. In these egalitarian communities, everyone is seen as pretty much equal. Giving out an award for the best tracker or hunter in a San group would be completely out of the question! Singling out any individual for heaps of praise would be seen as going against humanity's fundamental nature.

It certainly seemed to go against my nature. Many of the happiest moments of my life were when I was a child, wandering the seashore, finding treasures, and feeling absolutely safe and protected in my wild-world bubble. When I needed healing as an adult, I returned to the Seaforest.

I don't think any human being is truly prepared for what fame does to the psyche. And because I had never really sought out fame or recognition, let alone the kind that literally came overnight, the experience affected me profoundly.

I'd had these intimate, vulnerable—almost sacred—experiences and now they were being consumed by millions of people. It was as if my soul had been fractured into millions and millions of pieces that had floated into television sets all over the world, only none of those pieces were really *me*.

Yet, as much as I wanted to leave the spotlight behind, I also felt compelled to speak and lobby on behalf of the voiceless animals, for our great Mother. I felt caught between two powerful desires: my own need for quiet communion with wildness and a burning calling to awaken the love of nature in my human kin.

An Altered State

As the spotlight shone brighter, I found myself lying awake for hours each night, exhausted yet increasingly unable to sleep. I would stumble out of bed and pace around a home that I realized would seem completely alien to my ancestors even just a few generations back. My heightened senses would struggle to drown out the tame-world sounds and signals that had once been white noise—the low hum of the refrigerator, the kitchen lights that felt so cold compared to the feeling of sun on bare skin.

I was an animal whose natural habitat had been destroyed, and in its place was an artificial world built up to meet my every need and desire—and yet for all those comforts, I could not sleep.

A full four months went by during which I sometimes managed only ten minutes of sleep a night. Insomnia is terrifying, because without sleep, the mind just starts to disintegrate. At first, I bounced between jittery nights and irritable days, but after many weeks of this, I gave up all hope of sleep and began to experience altered states.

There is a reason sleep deprivation is a common torture method; when deprived of the basic need to sleep, people start to dissociate and hallucinate. It becomes hard to know what's real and what's not. Some nights I would drift off and manage to sleep for half an hour or so, but many more nights I was thrust into a nightmarish realm of darkness and strange sounds and visions. Sometimes I'd see the flashes of movement that reminded me of those strange figures I'd glimpsed in my room as a child, as memories of childhood fears got jumbled up with current-day paranoia.

The one saving grace was that I'd studied altered-state healing for years. I'd danced with San healers and gone into altered states of my own. I'd done visualization techniques during which it felt as though I'd left my body. I'd also studied the connection between rock art and altered states of consciousness with archaeologist Janette Deacon. There is evidence that the half animal, half human forms in this art depict the experiences people had in trance states. Geometric patterns similar to those seen in the first stages of trance experiences are also common in rock art.

So when I would see and hear strange creatures in the night, when the wall would melt into the kind of impossible geometry associated with altered states of consciousness, I was as aware as one possibly could be about what was happening.

In spite of all this, the suffering was profound. For months, I could hardly move properly, so I had to spend most of my time in our cottage while Swati took care of me, along with so many of our household responsibilities. I couldn't drive my car or ride a bicycle. And the daily dives that had once been so joyful and restorative had become grueling.

At the height of my tracking practice, I could hold my breath for a full five minutes underwater—even longer in perfectly calm waters. Now it was a struggle even to breathe properly *above* water. I once could stay out swimming in cold, deep water for two hours. Now I couldn't even last five minutes.

I felt such deep despair. It was as if I'd been granted a super-power, and had it taken away, leaving me weaker than I'd ever been. Physical and mental setbacks of any kind can be devastating, but they are uniquely challenging when they occur *after* you've made a conscious effort to turn your life around—like maybe nothing you do matters, like all your efforts to transform were just some cosmic joke.

I didn't feel like someone who'd worked hard to learn the language of nature. I felt like the scared boy I'd been so many years ago—the one who'd reached out to the creatures of the sea because they were a lot less frightening than the ones on land.

As a child, I'd been fearful of what hid in the darkness.

As an adult, I knew that what hid in the darkness was me.

Shattering

One morning the water temperature had dropped to 55 degrees, and I swam out into the Seaforest. As I pushed farther into the deep, I saw a bizarre sight: a three-spot swimming crab in the canopy of the kelp forest. This animal was completely out of its normal habitat of living underneath the sand on the seafloor.

I had seen such aberrations before—like the bat flapping around our boat in broad daylight the day of the crocodile dive in the Okavango Delta. This phenomenon—any disruption in the normal fabric of nature—was well known to the Indigenous trackers and shamans I'd worked with over the years. It was a portent that something powerful was about to happen.

At the same time my brain was telling me that the crab was out of place, my eyes couldn't register what was wrong with it. I was experiencing a kind of blindness to what was right in front of me in the natural world. My ability to recognize patterns in nature had become impaired.

After all the tracking expertise I'd gained over the years, it was a terrifying feeling.

This wasn't the only strange occurrence I remember from that bleak period of my life. My whole experience of the Seaforest seemed to be changing. Now when I went in the water I no longer saw the beautiful things I was so used to seeing. Instead I noticed dead carrier pigeons, litter, oil slicks.

The whole of nature was a mirror, and my psyche was shattering.

Temporary

During those long sleepless nights, I worried that I would never track the way I once had. I feared my health would never return to its robust level from just a year earlier.

I feared that my body and mind were irrevocably damaged.

Yet I still believed I could maintain a connection to the wild. I went in the sea every day for a few minutes at a time. Slowly, day by day, I increased the time, letting that wild system gradually bring me back.

One of the most disturbing feelings was not being able to

get enough air through my snorkel and having to keep my head above water to breathe. I was so weak I had to hold on to the kelp. I'd lost so much weight that none of my clothes fit, and I felt self-conscious with my shirt off. I'd always worked hard to keep my weight down, and it was strange to feel so thin.

Swati played a pivotal role in my recovery—when you're breaking apart, it's an amazing gift to have someone holding you together.

"Everything will be okay, Craig. This is temporary," she'd say, over and over again, her voice soothing. She never let me see that she was worried; she would just smile and hold me. She was so strong and confident that a small part of me believed her despite my shattered state.

I began working with a sleep scientist, Dr. Dale Rae, who suggested I try a regimen of breathing practices. I gave myself over to her care and did whatever she told me to do. One of the most helpful was a simple but powerful five-seven-five breathing practice. I'll share it here:

Breathe in through your nose, slowly and steadily, while silently counting to five.
Hold your breath for three seconds.
Breathe out, slowly and steadily, for five counts.
Repeat this five times.

The next time, breathe in for five counts as you did before.
Hold your breath for three counts.

Then breathe out for seven counts.
Repeat five times.

Then breathe in for seven, hold for three, and exhale for seven.
Repeat five times.

Breathe in for seven, hold for three, and exhale for five.
Repeat five times.

Finally, breathe in for five, hold for three, and exhale for five.
Repeat five times.

Then return to normal breathing, relaxed and easy.

All in all, the practice takes only about twelve minutes, and it's amazing how much it calmed me down every time. When I would wake in the middle of the night, my heart and mind racing, I would go through the breathing process and calm my whole nervous system down in minutes.

I owe my life to Dale.

While the elements of wildness I'd learned were important, they alone seemed insufficient to help me overcome this obstacle. No longer could I rely on some physical or mental challenge to bring me closer to health; no longer could I dive headfirst into danger as a form of salvation.

Instead, I had to learn the humbling practice of surrender.

I took a break from work, recommitting myself fully to nature. Each night, I continued my rituals: praying to my teachers, practicing my breathing. Each morning, I'd rise, make my tea, and walk down to the Seaforest. I made peace with the fact that the limits of today were not the same as the limits of yesterday.

Slowly, five minutes in the water became six. Six became seven. Seven became eight.

I surrendered myself to the recovery process, to the people caring for me, to life itself.

Breathe in, breathe out, breathe in again.

This act of surrender helped me bring a sense of spaciousness to this challenging period, to see it not as a permanent backslide but rather as a kind of rite of passage. As disorienting as the altered states were that I experienced during the height of my insomnia, they also seemed to grant me access to the inner workings of my child mind.

It was as if I had hacked my way into the matrix of my mind—and been given a glimpse into the way neuroses had taken root long ago. I realized I could heal those old fears and insecurities by showering them with forgiveness and understanding.

As I did, I began to feel my strength return.

What also helped my healing process was receiving thousands of messages from people of many different cultures around the world. All the stories told of a deep reconnection to nature inspired by our film. Several people told of being saved from suicide, of feeling uplifted, of having their lives transformed somehow. I was surprised and gratified, and these stories shone like beacons on those dark frightening nights.

My story no longer belonged to me alone, but was also gaining powerful resonance.

My soul was returning to me, stronger, healthier, less alone.

Returning

Swati and Pippa walk with me down the path to the Seaforest. As I enter the ocean, a deep joy suffuses my whole being. I marvel at the feeling of the water and the beauty of the algae. From the corner of my eye, I see four Cape clawless otters tracking my movements.

For some time now, these shy but playful animals have been my indicators and guides in the wild. I won't see them for months on end, and then in times of grace they appear. Otters always remind me of Charmaine and the mysterious experience we had in Pringle Bay years ago while performing the ceremony of calling to the water. In Africa otters are a potent spiritual symbol for many cultures because they move fluidly between the realms of land and water.

I catch glimpses of four bewhiskered heads and four long, thick tails. They approach from behind, as any good predator would, and then I have an otter swimming on each flank, as if they are escorting me back into my amphibious life. They are young animals, full of curiosity and energy. Then two of them abruptly swim off while two stay behind.

One in particular is bold and curious. He comes so close he begins to make contact with my feet. Pippa and Swati can't resist and join us in the water. The two otters swim around us, creating bubble trails, weaving loops of invisible magic around us. The boldest otter, who has a small wound on his hind leg, repeatedly touches our feet and legs, and then touches my face with his paw, which is so much like a hand with five fur-covered fingers. There is not the slightest aggression in his touch, just the purest curiosity.

For over ten minutes the otters play with us, an awe-inspiring experience. And then they leave, merging with the landscape like ghosts.

Afterward I feel buoyant, as if I am floating on air. I have given myself to the wild and she has responded in the most profound way.

Nature's Grace

We don't need to enter dangerous waters to be at the ocean's mercy. We may not always realize it, but we are only breathing because of nature's grace. Microscopic marine plants called phytoplankton convert the energy of the sun into the oxygen that makes life on Earth possible. The ocean keeps our climate stable and the earth cool.

What choice do we have but to surrender to this awesome power?

As for our primal fears, I don't believe we can chase them away or deny them any more than we can stop the waves from crashing against the shore or tell a cat not to hiss when her tail is caught under a branch.

But we can befriend these life-giving fears—know them, honor them, speak to them. This is a way of speaking across time to our ancestors, who knew perils we will never know but who also knew a kind of life that our tame world has tried to snuff out.

When we make this connection—when we realize that we are part of the wild, not separate from it—we come to see how much we *do* control. Wildness requires us to push up against our limits while never forgetting who is in charge. Nature needs two seemingly opposing things from us:

To accept that we are at the mercy of its awesome power.

And to rise to the challenge of protecting wildness at all costs.

CHAPTER SEVEN

CONNECT

JANNES, SWATI, AND I WERE DIVING IN SHALLOW WATER BY THE shore when we spotted something rarely seen in this part of the kelp forest: a southern right whale.

The water was cloudy that day, so only the whale's tail was visible while the rest of its body was obscured in the haze. We could still make out its great size: though still young, the creature was about thirty feet long.

What was the whale doing so close to shore?

A whale this young would still hold recent memories of being with its mother and perhaps whispering to her in the acoustic shadows. Perhaps it found the steady wave action so close to shore familiar, even comforting.

The three of us held still as the whale began to rub its tail flukes and the side of its body against the kelp. I couldn't see any obvious barnacles or whale lice it needed to remove, so I wondered what it was doing.

Could the whale simply be enjoying the tactile feel of the smooth kelp, not unlike the texture of another whale?

With the creature seemingly unaware of our presence, Jannes and I followed at a respectful distance as it moved away from the shallow forest into deeper water. It glided slowly and steadily until it encountered a large rock blocking its path.

Nearly all the whales I've encountered over the years have been very gentle, and none have seemed in any way intimidated by my presence. But today the water conditions were far from ideal—even less visibility here than in the shallow water. I also noticed scratches on the whale's back, perhaps the result of a recent run-in with a predator or boat.

The young whale may have felt trapped by the rock and got a fright seeing us, much as you or I might startle if a mouse or small snake brushed past our feet. All of these factors might help explain why it suddenly let out a tremendously loud screech that penetrated our bodies—the aquatic equivalent of an elephant trumpeting in your ear—and seconds later tried to crush me with a fierce smack of its tail.

I have no doubt the blow, had it struck, would have knocked me out or even killed me. Thankfully, it landed just to my side a few feet away. Still, the force of the movement alone was frightening, and the upsurge of water it created was enough to push me back a good twelve feet.

The whale swam around the rock and glided about fifty yards farther out.

Jannes swam over to Swati to make sure she was okay while I rested in the dense kelp, taking deep breaths to bring my heart rate down. I felt a little shocked, and sorry for frightening the whale, so I made several loud deep sounds underwater in the hopes of letting the whale know where I was and that I meant no harm.

I expected it to swim away, but a remarkable thing happened: the whale returned and positioned itself right next to me. To my amazement, it began to resume the tactile touching of the kelp, allowing me to get precious still images of this process. I continued to make soft vocalizations to give the animal a sense of my location. It stayed for about ten minutes, completely relaxed in my presence, before slowly moving away.

It was a deep privilege to witness this kelp-touching behavior, and I wondered if this had been seen or documented before.

But I was reminded that whales can be very dangerous, especially in such low visibility. If I ever find myself diving next to one again, I will make sure to give it space. If it comes close, I will immediately begin to make those gentle deep sounds to let the animal know where I am and that there is no reason to be afraid.

Tameness as Predator

Sometimes I'm asked if wild animals are growing more aggressive toward humans—if nature is rising up against humanity to punish us for our many crimes against our wild kin and our planet home. Recent reports of orcas ramming their bodies into yachts, for instance, have inspired people to question whether the animals are "taking revenge."[1]

While I understand how it might appear that these animals are seeking retaliation against the deadliest predator on the planet, I think that's a human interpretation of animal behavior. In fact, some recent science suggests that orcas are actually playing with the boats.[2]

I have certainly witnessed behavior in animals that could be called aggressive. We've seen it in Cape Town when normally shy Cape clawless otters began to approach humans, occasionally even biting them.

But what might seem at first glance like growing aggression is ongoing evidence of animals becoming more and more pressurized by human encroachment on their territory. As humans continue to invade and tame the last remaining wild places on Earth, we're threatening the very existence of many species—and that can push animals over the edge. Take the otters: many of them came farther inland than usual during South Africa's COVID lockdown, as there were fewer humans in the area. All that changed when restrictions were lifted and humans returned to the places the otters had previously had to themselves.

As gentle and nurturing as nature can be, it is also fierce. Animals hunt and kill. They will instinctively protect their homes, their food sources, and their offspring. Like humans, they also carry the memories of trauma, and when they reencounter the source of that trauma, they will naturally have a reaction.

But in my experience, the animals dubbed "killer whales" are gentle when it comes to humans. There are no instances of an orca killing a human in the wild.[3]

Of course, they have shown aggression toward humans in captivity.

These rare instances serve as a tragic reminder of what happens when tameness is enforced upon wild creatures, when animals are caged and confined and deprived of their freedom and independence.

The Distant Touch

The dangers of the tame world were much on my mind the day before Pippa's journey to Los Angeles—the place that in some ways is the very opposite of our world here on the tip of Africa. Though she was proud to be carrying the message of Mother Ocean at the Academy Awards, she admitted she was a bit frightened of having to go alone. She already felt disconnected from nature after more than six months of PR and pressure.

"What if I lose my wild thread, Craig?"

I understood her trepidation. I had decided not to attend the ceremony myself, as I was still working to regain my strength after my agonizing year of insomnia and anxiety. I was grateful to Pippa for embarking on this once-in-a-lifetime opportunity to speak for the creatures of the Seaforest, and hopeful that she and I would have an experience before she left that would help her hold on to her equilibrium and her innate wildness.

We slipped into the sea at one of our favorite spots, a wild bay at the base of a giant mountain. Clear water lay to our right, but my intuition told me we should go left, into the dark sediment-filled water. We swam for a quarter of a mile before finding ourselves in the midst of a shoal of about fifty smoothhound sharks. With long, slender silver bodies, some adults were longer than me, while others, just youngsters, were about four feet long.

I felt no threat from these predators. Their body language demonstrated no signs of agitation or aggression, just simple curiosity and caution. All thoughts of the tame world retreated, and I felt fully present among these graceful creatures.

Seeing so many large predators in such shallow water was deeply heartening, as predators are the sign of a healthy ecosystem. But why were they here?

Now that my tracking skills were beginning to sharpen once again, I had an inkling. At first, I considered that the sharks may have been hunting in the area using their superior electroreceptors—in other words, their ability to sense minute electric signatures of other animals. Their predation skills are further bolstered by receptors running along the length of their bodies that create a kind of pressure map of their surroundings. This ability to "see" without using their eyes is sometimes called "the distant touch."

But from months of tracking in this area, I suspected the sharks were hiding from a much bigger predator.

The Shark Hunters

Two days earlier I had watched two orcas heading toward this very bay. It was just before dark and my binoculars detected a glimpse of their distinctive black and white markings in very shallow water inside the kelp forest. I'd only seen these two from shore and didn't know them that well. Most wild orcas have perfectly upright dorsal fins, but this pair's fins were noticeably different. Possibly from some nutritional deficiency, one's fin flopped to the left and the other's to the right. The pair's hunting skills had given them something of a reputation, and they were known in the area as Port and Starboard.[4]

One orca went shallow to flush out the sharks, and the other went deep to pick up any escapees. I watched the shallow hunter surface and for a brief moment glimpsed the body of a shark writhing in its mouth. After that, the pair vanished.

An encounter months earlier gave further insight into what may have brought the smoothhound sharks to such shallow waters: Swati, Pippa, Tom, and I had found the carcasses of three sevengill sharks washed up on a remote beach. Upon close inspection, we found each was missing its liver. The shark fins exhibited the telltale scars of orca teeth—blunt, worn to nubs from biting the sandpaper-like skin of sharks.

I was fairly sure these sharks had been killed by Port and Starboard.

I visualized the attack: each orca grabs a shark by the opposite side fin and pulls with their prodigious strength. The poor shark literally splits in two, and the massive liver floats out and is consumed.

Even great whites are not safe from these specialist shark hunters. Orcas weigh some three times as much as a great white, and have a slightly faster burst swim speed, so the shark is almost helpless against this giant mammalian predator.

Though I knew that their death was necessary for the orcas' survival, I felt a kinship with the sevengill sharks because I've spent so many wonderful hours diving with these animals. To me they are like lions, almost the same size and weight as the big cats, and lightning fast. They hunt together and have large, powerful tails that deliver bursts of speed when needed. They can glide silently toward a seal or dolphin, giving off minimal pressure waves.

A silent attack ending in an explosive, deadly impact.

My most memorable dive was with a pack of thirty sharks. I had been diving with Tom, who was only eleven years old at the time. Imagine being inside a pride of thirty lions with a small child and feeling quite safe. I still remember so clearly these great shark bodies brushing past us. The water was crystal clear and everything around us was deeply peaceful, while on the surface a rainstorm was raging.

Ocean predators such as sharks are far more ancient than humans; they evolved hundreds of millions of years before we even appeared on the scene. Since this strange, ungainly swimming primate does not register as prey or predator, our presence generally does not trigger an attack or flee response. Sometimes curiosity is there, sometimes avoidance.

The idea of sharks as bloodthirsty killers just waiting to pounce on an unsuspecting human is most decidedly a fiction that prevents us from understanding their true nature. This fiction divides the world into wild and tame, with sharks on one side and humans on the other, when the truth is much more nuanced.

We All Know This Predator's Name

Thanks to several months of shark and orca tracking, I theorized that the smoothhounds Pippa and I found that day were hiding from the orca in a kind of acoustic shadow in the shallow waters near the shore. Orcas can sense sharks by echolocation, but the sounds and movements of breaking waves make this more difficult.

Pippa and I emerged from this exhilarating world and stepped onto land again, feeling the awkward weight of our two-leggedness, so stark after the gravity-free flight among those silver shadows.

We marveled at the grace and power we'd just witnessed. In 400 million years of evolution, these sharks have hardly changed. Their design is so immaculate and efficient that very little adaptation was required.

But now they need to adapt to a peril no sea animal has ever faced until recently, a creature that makes an orca seem like no threat at all. This superpredator is breeding like prey and has weaponry and an appetite that is deeply greedy and wasteful.[5] This creature has forgotten where it comes from, who it is, and why it's here.

I don't need to name this superpredator because we all know its name.

It's because of this predator that these beautiful sharks are ending up in Australia, marketed as fish and chips.

The marine biologist Sylvia Earle famously said, "You should be afraid if you are in the ocean and don't see sharks!"[6] We need sharks for a healthy ocean; they keep their prey strong and vigorous. The absence of sharks is ironically what is dangerous for us, yet sharks and rays are in serious trouble, as one third of this subclass of fish—known as elasmobranchii—is threatened with extinction.[7]

Before Pippa left for Los Angeles, we were reminded of the importance of telling these stories so that the members of our forgetful species can recall our wild origins and learn more about the magnificence of our wild kin.

Predators like sharks are the guardians of our seas—they regulate the ocean's ecosystem and are fundamental to its well-being. This means they are also the guardians of every breath we take. It is through their grace that we live and thrive. If we put that knowing into our hearts, we will do everything we can to care for them, love them, cherish them.

To cherish the creatures of the seas is to cherish ourselves.

Does Nature Need Us?

This earth-and-ocean planet is very old. Woven through it is a deep biological intelligence that keeps its highly complex ecosystems in exquisite balance. All the world's supercomputers combined cannot come close to matching the communication between multiple species of plants and animals, fungi, bacteria, and viruses. Our best space-age technology pales in comparison to the natural technology that allows an insect or bird to fly on a tiny drop of nectar fuel.

This ecological intelligence works hard to maintain itself, to create biodiversity and to keep temperature and humidity within life-supporting limits.

Yet our childlike species seems to forget we're woven into this delicate balance—and that is having a devastating impact on the biodiversity and life-sustaining functions that make our world habitable for ourselves and generations to come.

Business and industry leaders, in particular, and the officials who support them, seem to have forgotten that the bedrock of their establishments is actually nature itself, that she allows their companies to function. Obsessed with short-term profit, they consume and plunder irreplaceable resources, our planet's ancient forests and wetlands, its wild rivers and oceans. If biodiversity collapses, all their future investments will be worth nothing. As Swati's nature mentor Bittu Sahgal, the founder of the ecology magazine *Sanctuary Asia*, said, "Unless ecosystems are nurtured so as to return to their biodiverse character, plantations of millions, billions, or trillions of trees will not serve to rein in either climate change or the pandemics that were triggered by the human mismanagement of our planet's ecosystems." Our species must let nature regenerate so we can breathe and live from her biodiversity.

Human beings cannot live without the natural world.

But does nature need us?

While I was struggling to regain my health, I asked myself this question a lot. It was hard to see plastic ensnared in kelp leaves or read about extreme weather events and rising temperatures around the globe and not feel deep pessimism and despair. Anyone who cares about wild creatures has no doubt been haunted by some version of the thoughts I wrestled with on my hardest days.

If humans just went extinct, this whole planet would be revived. All of the ecosystems and animals would be much better off. Earth would thrive.

There's a lot of truth in that perspective, and it's hard not to look at the ongoing environmental destruction our species is unleashing on the planet and think, *Of course nature doesn't need us. It would be better without us.*

But I think there's a more complex answer to this big question.

Learning Regeneration

All the animals of the Seaforest—the fish and sharks, the limpets and bryozoans, even the orcas—eat the wind.

The wind, combined with the Coriolis effect of the turning Earth, pushes huge masses of warmer surface water offshore. This movement also draws the deep water into the shallows. When sunlight touches this deep, cold water, laden with nutrients, life blooms in a miracle cascade of trillions of phytoplankton—the foundation of the food chain.

One day I noticed that the wind had swept thousands of compass jellyfish up from the deep, vast pelagic zone into shallower waters. The jellies, bright-red glowing translucent lights, provided a hearty feast for the creatures of the Seaforest.

I watched as giant false plum anemones caught the jellies and gulped them down. Then three-spot swimming crabs emerged from the sand and began to feast on the fallen jellies. Even the sea urchins snagged the jellies and devoured them slowly with their long white pointed teeth.

Hottentot sea bream had a slightly different technique: they bit holes in the jellies. Four years earlier, I'd noticed these fish were after a delicious snack living in the jellies: hyperiid amphipods. These crustaceans live inside jellyfish and eat some of the jelly without unduly affecting their host. I wondered if the bream remembered this delicacy from so long ago, or perhaps the prawny delight was burned into their instinctive memories. But after close inspection of the jellies, I realized this year there were no amphipods present.

Then Jannes pointed out something remarkable: many of the jellies were repairing the big holes the bream had made in their bodies. I could clearly see the scar tissue where the holes had once been, and I realized this repair process was happening quickly, over just a few days.

I marveled at the regenerative power of these creatures—and at the way nature always seems to provide a mirror for my own human experience. I had been going through an intense healing and regeneration process as well; just a few months earlier I could barely hold my head under the water, but I was now able to hold my breath for several minutes again and endure the cold for almost as long as I once had.

I still don't understand the inner workings of this enigmatic mirror. What at first seems like a reflection has its own mysterious weight and gravity. I was, in essence, reaching through the glass and discovering that the barrier between nature and myself was just an illusion.

Seeing the regenerative power in these jellies—and in myself—gives me hope that our species can begin to regrow the parts of ourselves so many have lost. Through tracking, our senses and intuition sharpen. We are less seduced by the tame-world comforts that can be so easily bought and then thrown into a landfill. What we gain by repairing this thread will leave us feeling stronger, braver, more whole.

We need that feeling of wholeness both as individuals and as a species, and the stakes have never been higher. If we don't work together to repair this broken thread, I fear that our species may go the way of the spiny sea stars that have recently begun to congregate in False Bay. Their story is one we can observe almost as a fable.

A Tale of Balance

As summer approached, I noticed that the large spiny sea stars had had a bumper year. Where before I might have seen ten sea stars in a small patch of Seaforest, now I was seeing hundreds. Their broadcast spawning hit-or-miss strategy had produced countless masses. With thousands of spiny arms and tube feet moving in slow motion, they looked like a giant living carpet of yellow and orange. I was amazed to watch a wave of these creatures slowly moving across the seafloor, decimating everything in its path. Every one of them needs to eat to survive, and with so many stomachs to feed, it was carnage.

Years ago, I'd observed how the giant turbo sea snails snap their shells to throw off their attacking sea star predators—normally, defenses like this can keep the stars' predations in check.

But it was a different story this year.

The first few stars were thrown off by the turbo's defense, but eventually the snail tires. It closes its operculum, its trapdoor, allowing the attacking star to envelop it.

The star then takes its stomach outside of its body.

Imagine you are the turbo in your small house. You've locked the door. Just outside the door is a sea star stomach, and it has secreted acid all over your door.

You wait and wait, but you are getting tired and hungry and running out of oxygen. After a day or two, you can't hold out any longer. You open your door and the acid flows in and melts you, killing you, and the sea star sucks you up like a milkshake.

What a frightening predator! Thousands of marine animals in False Bay were being devoured by the advancing wall of sea stars. Even the clever octopuses were not immune; though a sea star cannot kill an octopus, it can cause a great deal of annoyance. I watched in awe as a young octopus tried to lift a huge sea star away from its den, trying and trying and eventually giving up.

Nothing seemed able to stop the deadly horde—except the miracle of nature's deep intelligence. I watched spellbound over days as the sea stars began to weaken and falter after having developed great lesions in their arms from a kind of natural wasting disease. In an attempt to soothe their lesions, they contorted into extraordinary shapes, placing their tube feet and arms on top of the lacerations.

But the work was done. Their numbers dwindled and the Seaforest biological intelligence found its balance once again.

Of course, I've made these poor sea stars the villain of my thought experiment, when they are no different from any other animal reaching toward life—reproducing, growing, feeding, and eventually dying. I've witnessed what happens when other species enjoy explosive breeding years, and I've witnessed the way the Seaforest pushes back to achieve balance.

The same thing will surely happen to us, on a much grander scale, because Big Mother is running the show, not us.

Is there a way to escape this fate?

A Paranoid Predator

I often ask myself why so many members of our species feel so hopeless about our future and disconnected from our past—and the answer invariably leads me back to the huge shock that took place 10,000 years ago.

For about 300,000 years, we lived as free hunter-gatherers. With the land and sea as our wild pantry, we hunted or picked our food as needed. There was no need to store wealth or provisions. We had tracking and foraging skills to sustain us, and hours every day to rest and play.

With the advent of the agricultural revolution, our world was turned upside down.

Our core security system—our connection to wild nature—

was severed. Insects and weather became enemies to conquer or control. No longer could we wander into wild nature and feast on the rich biodiversity. We had to wait for crops to harvest and defend livestock against predators. Our health deteriorated with less diversity in our food, and our brains and bodies functioned less well too.[8]

Make no mistake: the hunting and gathering nomadic lifestyle is a tough one, and it is easy to see why some people were keen to change to agriculture. When I've asked San elders who once lived nomadically which lifestyle they preferred, half of them yearned for the old way and the other half were happier with the comforts of the sedentary lifestyle. Both lifestyles have different benefits and downfalls—it's just that we've taken the tame to the extreme, and that is dangerous.

During this part of our human history, as food access was connected to planting and harvesting cycles, a kind of paranoia set in. We see evidence of deep fears around crop failure in many cultures' rituals and deities connected to planting and harvesting.[9]

The Industrial Revolution further distanced us from our food and water sources, and brought with it a relentless work ethic that didn't necessarily lead to a deeper sense of food security: hunter-gatherers used to work only about three hours a day to provide all they needed, but now people were working ten or more hours a day, with even children forced to work in dangerous jobs.[10]

Nature was no longer seen as our life-giving mother but rather as something to use and exploit. Our food and water quality deteriorated further with the growth of mechanized, chemical-based agriculture.

We are wild creatures designed to forage directly from Mother Earth. We have lost connection to our primary sense of stability and well-being, lost the skills that gave us confidence in our ability to feed ourselves and our families. It makes sense, then, that we'd feel the compulsion to accumulate as much wealth as possible.

Anything to overcome the feeling of disconnection, paranoia, and scarcity.

As we continue to undermine our wild origins and threaten our Earth's living systems, we erode the very fabric of who we are.

But it *is* possible to find our way back to wildness and freedom.

Light in Their Eyes

Just as our ancient forebears followed the rains, the flowering plants, and the herds, I continued to seek out the great modern trackers, following in their footsteps.

I was elated, one day, to get the opportunity to welcome the team of master navigator Nainoa Thompson to these shores. I was humbled to meet this crew of strong Polynesian men and women who'd crossed oceans without any modern navigational equipment. There was a light in their eyes—that is what it is to be fully human. In the lashing seas, which became their home for months on end, in the stars that guided their instrument-less journey, they had found deep meaning and love.

Together, we set off to the tip of Africa: Cape Point. The crew of the *Hōkūle`a* stayed close while I tracked in the intertidal area, showing them the bounty of a kelp forest ecosystem. We entered the cold water without wetsuits, and even though they came from a tropical island, they adapted quickly. They were fit and hardened by months of exposure to cutting wind and spray on the open ocean.

Nainoa shared with me the story of how his teacher, Papa Mau Piailug, began his own wayfinding training at age five. When the waves would come and lift the canoe, the boy would frequently become seasick. To help young Mau overcome this, his grandfather threw him overboard. While this may seem harsh, his grandfather knew it would connect the boy to the ocean, and provide future safety.[11]

Papa Mau shared this story with great love and respect for his grandfather.

"My grandfather tie my hands and throw me overboard the canoe, drag me behind the canoe . . . When I go in the ocean, I can go inside the wave. When I go inside the wave, I become the wave, and only when I become the wave do I become the navigator."[12]

Nainoa had lived so much of his life on the ocean that he had truly surrendered to his amphibious form. His deep love, the Pacific, is sixty million square miles of ocean—more than 30 percent of the Earth's surface![13] He calls it "the biggest country in the world."

The navigator has to find tiny islands floating in this great expanse without instruments, relying only on their intimate knowledge of the natural world.

"There is no separation from the bird, or the wave, or the cloud, or the lightning, or the rain, or the star, or the sun, or the moon, from who you are as navigator," Nainoa explained. "You are part of that completely. And you're only a navigator if you surrender to that."

His words were like a star guiding me in the dark: perhaps asking whether nature needed humans was the wrong question—and a denial of how we already are woven into the tapestry of life.

A Reciprocal Bond

One morning, as the drizzle turned to great sheets stirring up the sea, I sought refuge in a Stone Age cave overlooking the kelp forest where I had dived with Nainoa and the *Hōkūle'a* crew.

The cave was made of dune rock, sculpted by the elements, a wind- and rain-sheltered sanctuary like a large room. Rodent footprints and bat droppings showed me the story of the previous night.

As I stared out at the falling rain, my mind traveled north to the great rolling landscapes of the Northern Cape, where I first learned about rain from my friend Janette Deacon, the rock art specialist who is also a world authority on the Bleek and Lloyd Collection—eleven thousand pages of verbatim text from the /Xam San written down in the 1870s by Wilhelm Bleek and Lucy Lloyd.[14]

Though descendants of the /Xam speakers are alive and well all over the upper Karoo region of South Africa, the /Xam language is tragically now extinct. Most of the descendants don't know their ancestry because their history was not taught in schools and because of their status as farm laborers many of them did not attend school. Much of what we know about their teachings comes from the collection's pages.

Reciprocity with nature was hardwired into their culture. Nothing was taken from the wild that was not also restored. A /Xam man named //Kabbo described gathering the roots and stems of a particular plant while at the same time replanting a section of root in the hole he'd dug so the plant would grow and flower again. Another narrator described the practice of sharing meat with a lion. If someone came across an animal killed by a lion, they might take some of the meat but would always leave a portion behind for the lion: "Our parents used to say that we must not carry off all the meat, we must leave food for the lion at the place of the kill, we must cover it with the bushes on which we have laid the meat we cut, so that he finds it."[15]

For seven years off and on, I worked with Janette to photograph the rock art of the /Xam and tried to understand the people who once roamed this harsh, dry place.

What emerged was a picture of a community of people whose hearts were woven into the ecology in a reciprocal bond that may be hard for us to grasp today. Their art spoke of a deep love for the land and the animals and a powerful connection to rain and water.

My fascination with rock art led me to the home of another

researcher, archaeologist Renée Rust, who had painstakingly over years, standing for hours on rickety ladders, traced the most remarkable collection of ancient art.[16] On a big table inside her old farmhouse two hours east of Cape Town, she rolled out a massive tracing. At the sight, the hair on the back of my neck stood on end. I was lost for words as I gazed at the fifteen-hundred-year-old masterwork, an underwater scene depicting a community of ichthyoid therianthropes—half human, half fish forms.

Renée had interviewed Indigenous people who claimed to have seen these creatures called "watermeide"—or water maidens—in nearby water sources. They were believed to be connected to rain-making and to amphibious creatures, like otters and frogs, as well as to giant mythical water snakes.

While it all sounded very metaphysical, my body and mind were reacting in a powerful way.

I began to use my tracking practice of visualizing the past to understand what these images would have meant to the /Xam.

I imagined a group of people deeply connected to nature and particularly in love with water. I saw them collecting water, drinking water with gratitude, admiring the freshness and health of this life-giving substance. They began to see the water as a living entity, an intelligence, not unlike the way some scientists, like the forest ecologist Suzanne Simard, are seeing complex biological systems in nature. Simard discovered that trees in a forest communicate via an underground network of fungi, sharing resources, water, and information.[17]

I saw a marriage of the human mind and the intelligence of

water, followed by a coupling, a falling in love and a mating. From that fertilization a creature was born—the offspring of human and water, represented by a human body with a fish tail.

I'd found the elusive amphibious soul, depicted in the cave art of South Africa, close to my childhood ocean kingdom, and I'd read the words of people who had experienced these creatures firsthand. The emotional bond was so powerful that this hybrid creature appeared totally real in the world of these people.

These mythical creatures seemed to be the ancestors of the mermaids of Western myth and legend. They are like shadows of our deep past, memories of our ancient connection with water. Distorted through time and culture, heavily adapted for popcorn-strewn cinemas and C-grade television series, they remain popular because somewhere deep down in the psyche we revere the sacred marriage between human and water that has given us all life.

Pure Source

With the images of the vanished /Xam people floating through my mind, I sat beside a small brook that ran into the sea. The water was stained rust red from the tannins of the fynbos vegetation growing on its banks. As I drank from its pure source, it hit me that the water in this stream, the water in the clouds in the sky, and the water in the sea are all part of the ocean of water that's been here on Earth since the beginning.

My tiny mind had long separated the sea from the rivers and the sky, but now I saw it all as one great ocean, fresh and briny.

The rivers were the mothers of the salt ocean, and the clouds were the mothers of the rivers—all one moving mass of water, one great, planetary ocean.

In the grasses and the soil, I saw an ocean of liquid; in the trees and shrubs, slowly flowing oceans everywhere I looked.

A flock of sacred ibis glided overhead, and I felt the liquid sea in all the birds that ever lived, in all the elephants and snakes, orcas and smoothhounds, you and me.

A Messenger

Pippa returned from Los Angeles carrying the Oscar. Our film, *My Octopus Teacher*, had won the Academy Award for Best Documentary.

As grateful as we were for this unimagined success, the year of mad publicity had come with a price for Pippa, too.

"It wasn't just the trip," she explained. "It was a year of living in multiple time zones with no time to go into nature and longing for it all the time. The thing that had healed in me became severed and starved. After years of diving every single day and being in conversation with the kelp forest and the coast I was literally cut out of the nature club."

Much of her wildness had been leached from her being: she tired more quickly, couldn't last very long in the cold, and felt more anxious than usual. It would take many dives in the Seaforest to restore her.

We have so much to learn to balance the tame and the wild. How do we honor and nurture our wild hearts in the age of extreme tech and hyper-communication? We are like wild animals set loose in an alien world, unsure of how to survive, all our tracking cues sparking in weird ways. We are the lost species trying to find our original home, swimmers drowning in a sea of unknown liquid.

Cast adrift, holding tight to the faintest thread that can bring us to shore, we need to pull ever so gently to safeguard that lifeline to who we really are.

AROUND THIS TIME I GOT A CALL FROM MY FRIEND MATT. HE HAD been having a hard time with stress and health problems and was wondering if he could visit. As Swati and I walked along the water with Matt later that week, he shared that he'd recently survived a harrowing armed robbery. He was walking by a lake when a man had approached him, held a gun to his face, and then stole his things, leaving him shaking and terrified.

"I haven't been able to sleep," he said. "I just keep replaying the memory in my mind."

While Matt was sharing his story, something very strange happened. A crowned cormorant, a glossy black seabird with a small crest on its head, swam to a rock near us and regarded us intently with its reddish eyes. These coastal birds are normally super shy and avoid humans completely, yet it was obvious that this bird had no fear of humans. It seemed to be in perfect health. A ring on its leg suggested it might have been rescued and then released back into the wild.

Despite that, I wasn't prepared for what happened next: the bird flapped upward and flew straight into my arms.

I wondered if perhaps this was another instance of nature acting as a mirror, reflecting back to me the message I most needed to understand. In that moment, the bird seemed to represent my friend's vulnerability and his struggle to express the fears he'd been grappling with.

I also felt strongly that wild nature was crying out to her human kin, asking for our help.

Matt, Swati, and I were all stunned. After a few moments the bird flew to a nearby rock. As I considered the countless animal lives lost to human recklessness, overconsumption, and greed, I felt a sense of despair. Many people may care deeply for animals but are unaware of the horrors we are unleashing upon the natural world every day. I recently spoke with several marine biologists about the nightmare that is deep sea mining, an effort to extract metals from the seabed that will be devastating for the delicate ecosystems of the deep sea.[18] This irresponsible mining has the potential to destabilize the ocean and negatively affect all life on Earth.

This latest attack against wildness is even more appalling when one considers that many African cultures see the deep ocean as the place where their ancestors go after death. This view is not incompatible with scientific research that posits life on Earth originated at deep-sea hydrothermal vents 3.7 billion years ago.[19]

We are destroying the mysterious places from which so much

of life springs, desecrating humanity's most sacred burial ground and the original birthplace of our planet's first life-forms.

As I looked at the cormorant, my heart broke a little. Face-to-face with this beautiful diving bird, I saw wild nature reflected back at me in its eyes. Those eyes would haunt me for weeks.

Why We Are Here

Recently, Jannes and I kicked off a new project called 1,001 Seaforest Species whose goal is to document and share the stories and science of 1,001 species in our waters over the next five years. Swati chose the number, inspired by the classic Middle Eastern folktale collection *The Thousand and One Nights*, in which the queen Scheherazade tells her husband a story each night to save her own life. Her stories change the king's heart and mind, leading him to spare his wife and the lives of other women in his kingdom.

In the same way, we believe our stories will warm human hearts toward nature, moving them to protect the creatures of the Seaforest.

Jannes and I were still mulling over which species to include when, one morning, a young octopus hijacked my camera and turned the lens on Jannes and me. As we looked at the octopus' footage of these two *Homo sapiens* who have come to call the kelp forest home, one decision became clear: we would include human beings as the 1,001st species so we might never forget our rightful place in nature.

When I shared this with cosmologist Brian Swimme, he became very emotional.

I didn't have to ask him why he was so moved. This idea—that *we are the universe, inextricably woven into its fabric*—is one that he has spent his entire life and career trying to communicate. I'm continually shocked by how revolutionary this idea is to people, even brilliant scientists and ecologists, but I think it's become quite common for us human beings to see ourselves as separate from both our animal kin and the universe that created us.

But Brian sees things quite differently.

In *Cosmogenesis*, he shares thinking about the unique role human beings play in our universe from his mentor Thomas Berry—a Catholic priest, scholar of history and religion, and fierce environmental advocate:

> The early Earth in the form of molten rock gave rise to the atmosphere and oceans. Over another billion years of evolution, the atmosphere with the oceans, minerals, and sunlight gave rise to the biosphere with all its fantastic diversity of life-forms ... And now, out of that complex network of relationships, the Earth, through *Homo sapiens*, has come to know itself. That is why we are here.[20]

Inspired by this radical idea, I asked Brian what he thinks about the many coincidences I've experienced in my years of tracking—what I can only describe as a mysterious intelligence in nature that so frequently seems to mirror what is happening inside my mind.

"As much as possible, I refrain from thinking of people as individuals," he told me. "Each human is indeed a unique manifestation of the whole universe, and our uniqueness is profoundly significant, yes, but our unique individuality comprises less than one percent of our existence. What we've learned is that every being is a 'universe individual.' Each of us is an individual entity that is, simultaneously, both the whole vast universe and just a fragment of it." Within each one of us, he added, is a "vast tower of time containing all cosmic history."

He concluded, "We are awakening. We are discovering our place at the tip of a vast developing universe. And we know what to do."

This is why we have such an amazing role to play—not only to observe our precious world, and appreciate it, but also to help protect it.

We can rise to this once-in-a-lifetime challenge.

Beginnings and Endings

My time in the Seaforest has given me access to so much life. I've formed relationships with hundreds of animals, from those whose lives span mere days to those who can live for decades.

My time in nature has also brought me into near daily conversation with death. I don't think I've ever gone on a dive without witnessing one animal killing another. Over the last decade I've mourned the deaths of sharks and sea stars, anemones and amphipods. I've lost dear friends like my octopus teacher, and our cat Leon.

The day after Leon died, Swati and I moved around the house in a state of shock. We cooked breakfast on the fire and shared memories of our beloved cat, and shed more tears. Then we went for a walk along the coast. On a section of sandy beach, Swati's keen tracking eyes picked up insect bodies embedded in the sand, like tiny jewels.

I imagined a great wind blowing the insects out to sea and drowning them, before washing their bodies back to shore. We collected forty or so from a few square yards and placed them for a photograph. I found an angulate tortoise that had also been drowned by the previous day's huge seas, and a super klipfish that had been thrown out of the water and was drying in the hot sun. Fur seals and kelp gulls were playing in the surf. Seabirds were riding the air currents pushed up by the waves. Life and death were everywhere. We glimpsed the face of our cherished cat in all the death, and in all the life too.

It's impossible not to reflect on the differences between the way death is viewed in the tame world and the way it was regarded by those who lived fully wild.

As we modern humans moved further and further from our wild sources of food, we lost our firsthand connection to natural cycles of life and death. Of course, there are still many cultural practices that prepare us for death and grieving, but for people who foraged and hunted, death was a more commonplace occurrence.

Many of the shamans I met during my years of filmmaking had frequent experiences in trance states that gave them an understanding of death. They were also constantly in touch with the ancestors. So the idea that someone who died would simply vanish for good and not be present in any way would be very strange to them.

Prolonging life at all costs would also seem strange. If you belonged to a nomadic San community in hunting-and-gathering times and you were mortally injured, your family would build a little hut, serve you your last meal, and everyone would say good-bye. Then the hyenas or the lions would take you out mercifully within a few days.

I'm certainly grateful for modern advancements that have helped me heal from life-threatening malaria, tick bite fever, bilharzia, infected cuts, and even just tooth decay.

But I also am grateful for the deep perspective on death I've been granted access to by my human and animal teachers. I'm grateful for my singular experience as a human tasting wildness on our living, breathing planet. I'm grateful for the understanding that I am connected by a thread to every drop of water, every breath of air.

The Gentle Wild

A decade spent nurturing my amphibious soul has given me confidence in humanity's ability to build a society more in harmony with wild nature. This confidence stands in contrast to a lot of the conversation around the environment and sustainability, which seems to take as its premise that humans are essentially a "bad" species—that we are violent and selfish and that only the veneer of civilization keeps us from killing one another. The daily news certainly supports this idea that we are deeply flawed. Yet the science in many ways may point to something quite different.

While sometimes humans can be very cruel, we also have tremendous capacity for gentleness. I'm continually blown away by the compassion I meet in my fellow human beings, the desire to respond to the call of the wild with help.

While occasionally interventions can cause more harm than good, I am always heartened when I think about what a helpful species we can be. From the millions of shelter animals that are adopted each year to conservation efforts around the globe to urgent calls for ecological justice, so many human beings are recognizing we have an important role to play when it comes to answering the calls of our animal, plant, and elemental kin.

There's perhaps no better evidence that the thread to our wild ancestors is growing in strength.

Throughout all of the Middle Stone Age archaeology, there is little evidence that points to interpersonal or intergroup violence. About 70,000 years ago, there were probably fewer than 10,000 humans on Earth, mostly living in Africa.[21] Our deep origins—and the fact that we did not go extinct—point to our species being highly cooperative and altruistic. It's very powerful to realize that we have nonviolent origins and remained largely nonviolent for most of our time on Earth.

It was only with the advent of agriculture that the reciprocity with the wild we'd enjoyed for some 300,000 years began to break apart—and with it our psyches. The vibrant umbilical cord that connected human beings to the wild was severed, and this break has been traumatic.

And what happens when animals are traumatized? They often become violent.

I think back to my encounter with the whale, the way she lashed out when she felt trapped between us and a large rock. Or the angry jaguar who snarled when her tail got caught in a branch. I reflect on the aggression, the fear, the violence that is so prevalent in our world today.

I remember also how the whale returned to me after the initial shock and was gentle, almost apologetic. Or the way the jaguar's aggression turned to playfulness once her tail was freed.

Even under huge stress, humans, too, have the capacity to act from our gentle nature.[22]

Reciprocity

We live in a world that has experienced severe ecological destruction, yet life is just waiting to thrive if given the smallest chance. As we understand more about the climate crisis and the sixth mass extinction phase in our Earth's history—the fastest and most frightening, as well as the first ushered in by humankind—many of us are rethinking not only what we consume but also the way we live and work, how we structure our organizations and societies, what we value.

I recently chatted with Megan Biesele, an anthropologist who spent many years with the San and knows their language well, and Melissa Heckler, an author and researcher focusing on early childhood development. We discussed the importance of sharing in Ju/'hoan culture—a value that is instilled at birth when each newborn is given a piece of jewelry they are expected to pass on to someone else when they get older.

As I witnessed firsthand when filming with the San, the hunt is a communal activity in which the hunter is no more important than anyone else—in fact, it is the maker of the fatal arrow who owns the meat, not the hunter, and these skilled arrow makers are often women, the elderly, or people who are unable to hunt.

Reciprocity is even reflected in the Ju/'hoan language, in which the same word—*nlarohkxao*—is used for both teaching and learning, which not only respects a child's natural motivation to learn but also creates a balance of power between teacher and student.

"There is constant leveling," Melissa explained. "Maintaining the culture of equality requires constant vigilance."

If we can embrace the reciprocity that is part of our heritage—both individually and collectively—we will witness the resilience and inherent kindness of our species. We can heal this collective trauma by building a bridge back to source, back to our amphibious soul.

There really is no "other" in this world—whether animal, human, or plant, we all share the same air, same soil, same ocean. There may appear to be predators and prey and dangers lurking everywhere, but in the bigger picture there is tremendous support from the biological intelligence, keeping each living creature alive.

During a recent conversation, Nainoa told me, "We don't need hope, Craig. We need belief!" Belief, after all, is the basis for action. When we believe in our deep hearts that humans have the capacity to transform our ways of dealing with our earth-and-ocean home, we can draw on the wisdom of our ancient ancestors.

The well-being that results from this wisdom is something no tame-world comfort can bring. For most of our time on this planet—over 95 percent, in fact—our species lived as nomads. These ancestors of ours had no concept of "wild" or "wilderness." Everything was wild; there was no tame. Only when colonial powers brought the horrors of the tame to their world did they even consider the difference.

Caves and built shelters provided temporary protection and fire gave warmth. But our ancestors' deep security came from nature, because that's what they knew, that's how they breathed and survived.

Now our challenge is to strengthen the delicate threads to our deep ancestors. We can all find opportunities to rewild ourselves. I found my healing in the watery world. But wherever you are, nature is in and around you already, just waiting to help you flourish.

CHAPTER EIGHT

PLAY

If the birds up in the trees
Know how beautiful they are
If the mountains and the sea
Know how magical they are
If the stars which made our skin
Show how radiant they are
Won't they shine their light until
You remember who you are?

—Zolani Mahola, "Remember Who You Are"

EVEN AFTER MANY YEARS OF LIVING IN THE CAPE OF STORMS, FEED-
ing my amphibious soul daily, I still feel the pull of the tame
world. Some of this is my choosing; when writing, editing, or
responding to requests to help the Sea Change Project share its
message of deep personal ecology, I can spend long stretches of
time at the computer without coming up for air.

The technological seduction is strong and I'm often drawn in without thinking. Of course, these habits are addictive by design. Some days it seems the giant tech mind wants nothing less than to suck my amphibious soul dry. It wants me docile, my attention closed and fixed: small world, small mind.

After a long day of editing, I sometimes fantasize about going fully feral—deactivating my email account, closing the door of my editing studio, and finding a remote island where Swati and I can immerse ourselves even more deeply in nature. I have the survival skills and experience to do it: when I lived on a tropical island for half a year in my twenties, my only convenience was a tiny tent to keep the mosquitos at bay.

It was idyllic in many ways, but eventually I felt the great pull back to Africa, the mother continent of our species, eager for what she would teach me. After more recent trips to remote locations where few humans go, I felt an equally strong pull back to family, friends, and tracking partners. We are a social species, and while I enjoy my alone time with nature, sharing wildness with friends both new and old has been one of the most enriching parts of my journey.

I am also careful not to write off every aspect of the modern world. I need only look at the advancements in filmmaking over the last three decades to be reminded of how the wildness of human spirit continues to transform the way we tell stories, make art, connect with one another, learn. My morning cup of tea, my afternoon soak in the sauna, my ability to chat with friends across the globe in the blink of an eye—when I'm fully awake, I can look at every one of these experiences with the same sense of wonder and gratitude I bring to my morning dives.

Besides, it's much easier to immerse myself in cold waters, endure biting insects and the inevitable bumps and bruises, when I know I will return to a warm, comfortable house once the adventure is over. It's a whole different story to try to open the door back to a hunter-gatherer lifestyle. In many ways that wild door is closed and locked. When we stepped through from our foraging past to a world built around controlled agriculture, we didn't realize that the journey was one way, no going back.

I wonder: If we somehow lost all our comforts and were transported back into a nature paradise, would life really feel all that enchanted? Or would we continually yearn to watch a film, read a book, taste the perfect cup of coffee—not to mention enjoy the convenience of modern medicine?

We are a conflicted species, part wild, part tame, and hungry for the best of both worlds. The problem is these worlds don't support each other. What's more, one of the most nefarious aspects of the tame world is the way it has systematically cut large groups of people off from everything that heals, sustains, and nurtures life.

My time spent living in an African township was a brutal reminder of how the comforts of privilege are the direct result of the oppression of so many others.

The way forward for the amphibious soul whose heart beats for every living creature on this planet is not always clear. We cannot simply hop into a time machine and "redo" colonialism and the agricultural and industrial revolutions more mindfully or equitably. We can't all move into the wilderness because there isn't enough wilderness left.

But too many of us are hungry for a different way of living for us to give ourselves entirely over to the tame.

Perhaps the next stage of evolution of the amphibious soul means moving fluidly between the world we were born into and the world we are creating. Creatively, lovingly, communally, and—perhaps most importantly—playfully.

A Soul Remembering

One afternoon, after I had just finished a long day of interviews, a deep fatigue set into my bones. Part of me was ready to call it a day, but Pippa convinced me a swim would set me up for a much better night's sleep.

As we were about to enter the water, we encountered a group of young men and women. One introduced himself as Shayan and said he was from Pakistan, traveling with friends from Zimbabwe and the Congo. They were curious about what we were doing.

After such a tough day, I wasn't feeling very social, but I noticed the way they kept glancing at the sea and the small waves breaking on the shore. After chatting for a few minutes, it came out that none of them had ever been in the ocean or knew how to swim.

Shayan and his friends were all terrified of the water, yet attracted to it at the same time. As they gazed longingly at the waves, there was a tension in their bodies, and they admitted they were afraid of drowning.

While many people have an instinctive fear of water, curiosity also pulls us toward its shiny, rippling surface.[1] We know it's potentially dangerous but also brimming with joy and healing. I offered to take them into the sea.

"It's not deep where the waves are breaking," Pippa said. "And we can hold your hands."

They were keen but still doubtful.

"It's okay. We'll look after you," I reassured them, feeling the responsibility of my words. "We know this sea—we go in every day. It's not strong today. We won't go far. Just relax your whole body, and breathe through your nose."

Holding hands, we advanced into the gentle, rolling waves.

Immediately, the little waves knocked all three of them off their feet, because they were completely unused to its thick, moving liquid force. When I bent down to reassure them, I saw that I didn't need to—they were laughing, entranced by the bubbles and the movement. Pippa and I looked at each other, matching grins on our faces.

This first encounter with the ocean is like a soul remembering its amphibious nature. My time with my new friends transformed my mood from exhaustion to ecstasy in a matter of seconds. My desire for solitude had vanished. We played like exuberant children, jumping and laughing and splashing one another.

"It's so salty!" Shayan said, wiping the water out of his eyes before the next wave hit him.

Though we were not fluent in one another's native languages—English, Punjabi, French, and Shona—we could all speak the primal language of joy. Nothing else needed to be said. It was a privilege to experience these first moments of salt and spray and ocean wonder with these souls. I'll never forget it.

Playing in Nature

Time in the wild reactivates the child mind, reminding us that we are both adult and child in spirit and that wonder is all around us. The stress we feel as adults to achieve certain things or to behave in a certain way puts the brakes on our curiosity and wonder— behaviors that are quite natural in children, both human and animal. Responsibilities born not out of joy but out of anxiety and expectation separate us from our wildness. But there is no more effective way to escape the world of the tame than to spontaneously play in nature.

While it is always joyful to share the Seaforest with friends, I look for ways to play even on my solitary swims. Bodysurfing is a very intimate way to feel the power and joy of the water. No equipment is needed.

Standing up to my neck in the ocean, I push off the sandy seabed just as the wave begins to break, and with three hard, fast strokes, I match the speed of the wave. My body becomes a living surfboard and glides on the surface of the wave over 150 feet toward the shore.

One morning I descended the winding path that overlooks the bay. The water was rough that day, but a small sandy beach in between giant granite boulders provided a relatively safe place to fly.

Choosing the right waves is critical here because of the steep shorebreak, which will dump me into the shallow surf, breaking bones and scouring skin. I've taken enough beatings over the years to know to look for the smaller, flatter profile waves.

The heaving swell sent me wave after wave, a bubbling, salty champagne. Even the swim back out into deeper water was thrilling; as I ducked under the waves, I felt their strength.

For my next ride, I tried another technique. Instead of catching the wave on the surface, I dove down, waited for the wave to pass over me, then kicked hard off the bottom from just behind the break. The wave had a strong pull, and this force sucked me from behind and then shot me like an arrow through the wave and out the front.

After half an hour, I'd caught about twenty waves. My body felt strong, my mind open. I looked again at the large rocks on either side of the beach and realized they were shaped like anvils. I imagined thousands upon thousands of years of moving water and sand slowly wearing away the sides and forming these shapes.

This was a track made by the ocean itself.

The same force that allowed me to ride the waves had crafted this ever-changing sculpture. I looked around and saw that force everywhere, written in the rocks, a language of smooth, rounded forms mirroring the flow of water and sand.

As I walked up the beach, I spotted a speckled Cape skink in a pile of dried kelp, hunting the small invertebrates that feed on the kelp. I saw the entry tracks amphipods leave when they burrow into the ground after their night feeding. I saw the feeding tracks of brown ibis, pecking holes to trap amphipods: their large beak impressions fill with the crustaceans, allowing the birds to pluck them out one by one, like using chopsticks to eat from a tiny bowl.

All around me I saw stories in the signs I've taught myself to read over the years. I felt totally free from time and body, just pure mind searching for the spirit of wild things.

I took one final dive into the surf as the last light silvered the water's surface. I flew down the shining waves breaking both left and right, and somehow the energy in the water transferred to my body. I emerged touched by the sheer primal joy of feeding the amphibious soul the food it craves: big slices of silver water served on a bed of frothing ocean.

Awakening Wonder

We can all develop a more playful relationship with nature, whether that means collecting crisp leaves or smooth rocks to use in our artwork or watching the squirrel perform acrobatics outside our window. While it may not feel instinctual at first, the more time you spend in the wild, the more your sense of wonder will reawaken. Your eye will see a digging implement where before it saw only a broken branch. A daily walk that once felt like obligatory exercise will become an indispensable respite from the tame world of expectation and responsibility.

Having a firm sense of my limits and knowledge about whatever ecosystem I'm exploring helps me play with my edge. The more I know the ocean, the more cautious I am, but also the more playful and adventurous I can afford to be.

One of my favorite spots to play is a place I call Cannon Rock.

When the waves hit this huge rock, it sounds like an old cannon firing a shell in slow motion, and the water explodes upward like a cannonball hitting the sand. It's lethal in certain conditions, but quite safe and spectacular in others.

One morning Jannes and I set out toward the great rock, swimming hard through that day's rough waters. I've grown accustomed to the murk of the kelp forest during low tide, and it's become a kind of meditation to watch my hands drag backward, tiny eddies of air collecting behind each finger. I used to feel a bit nervous in these conditions because it's hard to see predators, but now I trust the sea and my ocean routes. I've also learned to subtly use kelp and rocks as protection and to avoid swimming in these conditions near "hot spots" that are favored by bigger sharks for sneak attacks on seals.

When we finally reached Cannon Rock, a fair-size swell moved in fast. I moved close to the rock, which would normally mean suicide. The mass of water reared up like a giant horse, lifting me high up next to the rock face, then it curled back and the salt water rained down upon me in a heavy shower.

Jannes cried, "Are you all right?"

From where he was standing and filming, it looked like I had been crushed between rock and wave.

Normally rocks and big waves are a recipe for disaster, but I had observed this area closely for years and knew how to play in it safely. The power of the waves hitting the rocks and transforming into white sheets and tunnels of water sent energy zooming through my body and mind.

Just before we turned to swim back, a humpback whale soared out of the water, coming back down with a tremendous splash.

Sharing Wild

I'd always hoped my son, Tom, would view the natural world with the same sense of wonder and playfulness my grandmother and great-grandmother had encouraged in me. When he was little, I told him a story every night before he went to sleep. I spun elaborate tales about supernatural creatures and humans with special abilities to track and communicate with wildness. One of our favorites was about the huge hairy men twelve feet tall who came to a village to find a special boy called Tom Braden Peace.

His parents allow him to live in the forest with these men, who teach him how to speak with animals. A spider who lives on his body weaves him a cloak of silk. A snake and a scorpion live in the boy's silk cloak, and a fish teaches him to breathe underwater. He learns insect and bird language as well as the ways of the creatures who live underground. The boy has many death-defying adventures and eventually saves his family by using the skills of the wild.

Of course, the story was a fantastical projection of my own mind, a reflection of my deep desire to understand my origins, and to pass this knowledge to my son.

But I tried to make these stories real by inviting Tom into the world that had nurtured and sustained me all these years.

We would play in nature sometimes for eight hours straight, creating games using found objects. We would pour dry sand into water to see what creatures we could make, or balance rock upon rock, making impossibly high towers.

We wrestled and did rugby scrumming on the beach, churning up the sand into fantastic tracks. We wrestled underwater, holding our breath. This developed his balance and strength, but also his trust in his own limits: he learned never to push too hard, to bring gentleness even to wild play.

As Tom grew stronger, it became more difficult for me to keep up. I taught him boxing, but that ended when he knocked me clean out for a few seconds. A little blood, but no harm done.

In time, we progressed to flying objects, throwing abalone shells into strong winds, then watching as the wind whipped them back to us. One spinning shell stayed airborne for many seconds before it shot back to me like a boomerang. We built small dams to block rivers, and used kelp stipes as woodwind instruments, transforming shells and bones into drums.

Some of my most cherished moments have been with Tom, lost in play and laughter.

Curious Creatures

Occasionally while we were playing in nature, animals would approach Tom. My clearest memory is of two remarkable encounters with baboons.

The first was when Tom was about seven, and we'd just hiked up a river to a waterfall. Tom and I were lying on a rock in the sun, playing a game of balancing long sticks and branches, when we were approached by several baboons. Tom had seen these primates often, and had little fear of them, and I advised him to relax and see what they would do. Three of the baboons began to groom him, tugging on his hair and clothes in the same playful way they do with one another.

On another occasion, Swati, Tom, and I were sitting on a cliff overlooking the sea where baboon troops forage in the intertidal zone for shellfish. A troop of about twenty approached, and a huge male walked right up to Tom and me.

I told Tom not to look him in the eye, so as not to challenge his dominance. Tom was wearing a hat with a brim, so he pulled it low and looked down. To my amazement, a younger baboon walked up to Tom, lifted his hat, and bent down to gaze into Tom's face. Tom kept a straight face, but Swati and I couldn't help ourselves, and we both burst out laughing.

IT'S VERY RARE FOR WILD ANIMALS TO SEEK OUT HUMAN CONTACT, so when they do, the images burn deep into my memory. One very strange day stands out in my mind.

It was a morning of bright, hard sun, with a gentle southerly breeze, and on the edge of the low tide cycle. In the small bay ringed by kelp, the ocean was flat with a few ripples, the waves breaking and calming on the edges of the kelp forest about five hundred feet out.

The water was barely three feet deep here, and the ocean bed was a medley of seaweed and algae, ranging in color from light green to deep browns and rust red. I spotted a colorful assortment of mullet, sea bream, zebra fish, dreamfish, and of course klipfish, a staple on any dive. It is not unusual for these intelligent and curious fish to swim close or even follow humans around. They dart out from behind kelp and seaweeds, come close, then scurry away. Sometimes they follow at a distance just beyond the length of my arm and have a way of looking at me with their big protuberant eyes that makes them appear eager to engage.

As Swati, Pippa, and I swam out toward the kelp forest, some klipfish shot toward us, several of them bumping into our masks. While this has happened before in very shallow waters, there was a boldness to these fish I found intriguing.

After exploring in the kelp for a while, I decided to return to the shallows to take a closer look at the beauty of the variegated ocean bed.

Heading back, I could feel the low tide flowing toward high tide. As it was the day of a full moon, the low tide would be extra low, and the high extra high. As I swam in, I felt something soft bump against my torso. I assumed it was broken kelp floating in the forest, but at one insistent tug I caught a super klipfish nibbling on my toe.

What followed astonished me. By now I was in water barely two feet deep and floating horizontally, staying as still as I could by holding on to a rock. Two super klipfish with their big headlamp eyes and fantastic patterns hovered just a foot from my face.

I held tighter to the rock, relaxed my muscles, and tried to keep my energy calm and unthreatening. Within minutes I was surrounded by at least thirty klipfish, resting on my hand on the rock, nipping my bare stomach, darting up to my face, even nibbling on my under lip.

These fish were not accidentally bumping into me; they were choosing to swim up and actively make contact though I had no bait or food. Slowly I held my other hand out, palm facing up, and within seconds fish were swimming into my hand, resting on their side on my palm, nibbling my fingers. Their numbers kept growing. At one point I had four fish piled up sideways on one another in their mating postures lying in my hand.

In my thousands of encounters with wild creatures, I can certainly say that never before had any copulated on my palm!

I called to Swati and Pippa, who had been observing from the shore. As they swam toward me, I thought surely now the fish would get nervous, but instead their numbers grew. Soon each of us was surrounded by two varieties of klipfish: the agile and the super.

One stayed in my hand for several minutes, happy to be held and stroked. He even allowed himself to be lifted clean out of the water a few times before deciding he had had enough. I then had an assembly line of fish swimming into my palm to be tickled for a few seconds before each one would leave and another would take its place.

This was a remote area where people hardly ever dive, so this behavior was deeply mysterious. What made the klipfish cross that invisible line, that mutual wariness that usually separates humans from wild creatures?

We are enormous creatures in their environment, and yet the klipfish chose to swim past that barrier and take a risk. I don't know what they got out of it, but I do know that it gave me a sense of exhilaration, the kind of deep joy that only comes when a wild creature chooses to share its space and trust.

Such moments offer a glimpse of grace and the acknowledgment of the wild in us.

Though we returned several times to the exact same place, not a single fish ever behaved like this again.

After more than ten years of diving every day, the ocean keeps teaching, showing me more and more wondrous things.

Human life is not unlike the ocean, cycling through its waves and calm and storms, with times of plenty, times of hardship, times of pain.

Healing Fictions

Once our basic needs are met, finding purpose in life is critical for getting out of bed. In dark times, I've felt no purpose or energy to do anything, just a scary void. And that feeling has weighed heavily on my loved ones. There were days during my recent bout of stress and insomnia when I wondered if I'd ever recover my former strength.

But thankfully nature has a way of rejuvenating the soul and by some miracle this thing called purpose comes back. I feel reconnected to my desire to share the creatures of the Seaforest with the world, and then I can get up very early in the morning, full of energy and drive. Where this energy comes from is a mystery; my friend and crocodile dive partner Roger Horrocks calls it our "healing fiction"—a story we make up to feel good about our lives.

Sometimes I play with the idea that life is totally purposeless; it's easy to make that story work, and it can even be quite freeing at times.

Why not have these two opposing ideas live side by side? One, the idea that I have a deep purpose in life, to explore and share all things amphibious. And two, the idea that I am just a speck in an infinite universe whose actions only mean whatever story I'm telling myself that day.

Whenever I find myself clinging to either of these extremes too tightly, I ask myself if there's a better story to be found in the in-between.

WHETHER THIS SENSE OF PURPOSE IS ONE I'VE MADE UP OR IT rises from deep within, I cannot deny the healing powers of the wild.

One rainy day as I bicycled back from the shore just east of my house, my front wheel dipped into a hole in the road and I flew over the handlebars onto the road. The pain was immediate, electric, everywhere. But my shoulder bore the brunt of it—I'd managed to snap all of the ligaments that attached my collarbone to my shoulder blade. A specialist advised against an operation and instead recommended letting the scar tissue and muscle re-place the ligaments. I was in serious pain for days, and it would take over a year to get any strength back in my shoulder.

I was determined to keep going into the water wearing a sling, but swimming with one arm wasn't ideal. At the time, I was also nursing other maladies: plantar fasciitis, a searing pain in my foot; and my chronic surfer's ear.

This was how I found myself confined to a narrow tidal pool about four feet deep one afternoon, while the rest of my team at the Sea Change Project headed for a beautiful barrier reef. As I imagined the wonders they'd encounter, I couldn't help feeling sorry for myself. I'd trained so hard to get fit and strong again, and now my strength was slowly draining away once more, barring me from so many of my activities, like swimming, bodysurfing, and kayaking.

Just then, a flash of reddish brown caught my eye: an octopus I'd seen a few times, hiding in her den.

After a short while, she emerged from her home and glided along the natural rock wall into very shallow water. As I followed, she led me into a hidden crevice near the surface: a cave encrusted in glowing pink coralline algae. The cave was small but easily fit her body, and the opening was big enough for me to look in from outside.

The octopus moved near the surface of the cave, and the still water of the tidal pool created a perfect mirror, reflecting two octopuses back at me. My mask was just below the surface and my eyes at just the right angle and depth to create a silver mirror that illuminated the octopus and her reflection. What I saw was mesmerizing. I was inside an octopus dream, looking through a kaleidoscope, as the animal morphed into increasingly fantastic reflective shapes. As her arms moved, the reflection moved in symphony, until the reflection and the actual animal blended into one singular cephalopod mandala. No longer was I stuck in a shallow tidal pool with multiple injuries: I was face-to-face with a natural mystery beyond my rational mind's understanding.

I was on fire, my physical pain forgotten.

The kaleidoscope reminded me of another octopus encounter—when I watched one camouflage itself in two distinct colors, split right down the middle, which suggests the octopus brain works in two halves. It seemed to me a metaphor for the different selves we all possess, and how extreme experiences—injuries, illnesses, brushes with danger—have a way of bringing us face-to-face with neurotic tendencies that might otherwise remain just below the surface of our awareness.

Sometimes encounters with our mortality can create great suffering and confusion. But if we can face our fears and vulnerabilities in the mirror of raw nature, healing and integration can unfold.

Song Catching

Whether I'm trying to heal from an injury or lift my spirits, it always helps to bring a sense of playfulness to my tracking practice.

I'm interested in all kinds of tracking, but one of the most intriguing ideas was introduced to me by my friend Jon Young, a renowned tracker from the US. It's called "song catching," a practice Jon learned from Bill Monroe, who was considered the father of Kentucky bluegrass music.[2]

In the 1970s, Bill and Jon used to spend hours rambling in the woods while opening themselves and their intuition to the music of nature. In flashes of inspiration, they would hear the songs of trees and animals and wild places, and then "catch" them, scribbling down on paper what they'd heard before it vanished back into the ether.

Song catching is, of course, much older than Bill Monroe. Indigenous people, like the San shamans I'd met in Namibia, have caught songs from the beginning of time. My sense is that song catching may have its origins in animal tracking. Our early ancestors were deep inside nature and constantly heard the sounds of animals, insects, wind, and storms. One day someone imagined the track so skillfully, they left their body and became the animal. They'd invented trance—the song of the great healing dance.

The mother of all music and dance, this kind of trance involves rhythmic singing and clapping, and a technique of hyperventilation. There are many San language groups with different customs but they practice a similar form of healing dance with breathing and sound—participants enter an expanded state of awareness that allows for healing to take place and for the community to bond through a shared sense of oneness.

Just as trackers tried to catch the songs of birds or big cats, song catching can be done for a place. When the human mind merges with the biological intelligence of an ecosystem, in that meeting and expansion of consciousness, the song is born.

With these ideas in mind, I set out to try to catch the song of the Great African Seaforest.

It was all very well for Bill and Jon to catch songs, because they were accomplished musicians and songwriters. I, on the other hand, had never written a song in my life. I had no musical training apart from working with the musicians and composers who created soundtracks for films, and my sense of rhythm and melody left much to be desired.

For weeks, I went into the Seaforest every day and opened myself up, expecting a flood of profound words to come tumbling out in perfect song form.

And for weeks, nothing happened. I tried harder and harder, and still, nothing came. I heard the same comforting sounds I had heard each day and had come to call home, but nothing sounded remotely like music.

Playing the Water

At the same time I was struggling to catch the sound of the Seaforest, my son's musical gifts were blossoming. Tom has a special ear for music, and since he was young, he would play wild tunes on the organic pieces of flotsam we'd gather on our tracking adventures.

Eventually he learned to play the water itself, something I'd seen done in Central Africa. The flowing water of a stream or rock pool becomes a living skin that flows smoothly between each cupped-hand strike. The water drums sounded magnificent, and we had recorded Tom playing them.

This nature-based creativity requires making something from relatively little—or, in Tom's case, *hearing* something. This play embedded in Tom a kind of creativity that no human toys can teach, wiring the brain for deep, creative, multidimensional thinking.[3] It is a kind of tracking that may even help skills like math and physics. But more important than its impact on academic achievements, nature play develops calmness and confidence as well as humility in the face of the vast natural world.

Eventually Tom took his skills of math and play and woodworking into another realm. He began a project during COVID lockdown that would test both of us to our limits: he set out to build a full acoustic drum kit in my garage using only very basic tools.

I thought it was impossible and told him so, but he refused to give up despite multiple setbacks along the way.

He had to design and measure a series of straight pieces of wood that would somehow fit together into perfect circles upon which the drum skins would rest and resonate. The construction had to be accurate within two millimeters, something I could not imagine achieving in our rudimentary workshop.

And yet the challenge did not deter him; he crafted special tools and gigs using pieces of old metal and wood I'd collected. For months, the entire garage was covered in a mountain of sawdust.

Some days I'd have to sit for hours rotating the drum while he cut round and round with a router, only to have it break at day's end. I was sleeping badly at the time, and the process was painful. I often dreaded these long sessions because I was so tired and the dust was affecting my lungs. Still, I realized this was a kind of deep initiation for him, and for me.

One year later, Tom had a perfect full set of five drums. He could play this kit that had pushed us to our limits and had tested his workmanship and his tenacity. I was so proud of him, and every time those crazy loud drum sounds shook the whole house, I would just lie back and bathe in the sound of his triumph.

The Essence of the Place

Not long after I'd embarked on my song-catching adventure, I received a serendipitous call from Yo-Yo Ma's foundation. The great cellist was coming to South Africa and had offered to help with a music project. On a whim, I asked him to help me with this new tracking mystery.

Now the pressure to find the song of the Seaforest had intensified. Once again, I went outside hoping to hear words and melodies, and once again I came back no closer to having captured any songs.

Desperate for guidance, I reached out to Jon and to Anna Breytenbach, a professional animal communicator. Their advice made it clear that I'd been trying too hard to look for something specific—to catch the song in the same way Jon had done.

What they were telling me was "Don't try catching a song, try catching the *essence* of the place."

With their words in mind, I adjusted my approach. I reached out to the ancient biological intelligence of the Seaforest much like I did when I first began my daily dives. I asked the ocean to share her song, and then tried to connect my mind to those of my recent ancestors as well as the very old ones who had lived on this coast.

Finally I began to get snatches of words and sentences, more poetry than song. Some were sweet, others powerful but strange and not at all what I'd expected.

Then I was struck with an even more powerful realization: I'd been trying to song catch on my own, instead of asking friends for help.

Only when I brought my friends into the process did the real magic start to show itself, proving big hive mind is much more powerful than my little individual mind.

I enlisted the help of Jannes and Faine Loubser, the daughter of one of my closest friends from high school. Faine is a modern version of a Viking shield warrior—six feet tall and very strong but with the gentlest of souls. Like me, Faine is a natural scavenger, a lover of artifacts and bones and strange objects brought in by the great tides.

On one of their first song-catching missions, Jannes and Faine discovered a huge natural drum, a five-ton boulder that could be rocked back and forth, making a resounding thump. Next, Tom and I began to experiment with turbo and abalone shells that made a sound like popping water.

We also looked at the collection of things I'd scavenged from the Seaforest over the years and asked if there was anything to be done with them. When I mentioned to Jon that I'd found a whale ear bone years earlier, he was hopeful we might use it in the song.

The problem was: as hard as we tried to play the bone, no one could get a decent sound from it.

Then, inspiration struck: we decided to free dive into an underwater cave and play it there—even if the sound was inaudible, we loved the symbolic nature of the mission. The cave we had in mind is a pyramid-shaped underwater room about fifteen by thirty feet with a hole in the ceiling where a shaft of light pours through.

Faine, Pippa, and I took deep breaths, then dove in, and when we got to the bottom of the cave, Faine struck the whale ear bone.

I was stunned by what happened next: a small pocket of air trapped in the ear bone underwater helped to create a deep resonant boom.

For maybe a hundred years this whale had used its ear bone to listen to the song of its kin and thousands of other ocean sounds and now we were using it to send sound waves into the ocean around us. The vibration was so intense Pippa and I could feel it inside our chests like a second heartbeat.

We broke the surface shouting with the joy and wonder of the moment, and Carina, the Sea Change Project's executive director, shared that she'd even heard the sound on land. Again and again we dove into the cave, experiencing this extraordinary sound, which changed in pitch according to the amount of air trapped in the whale ear bone.

All in all, we discovered over twenty instruments made from giant Seaforest shells, dried and hatched shark eggs, kelp, and skeletons of animals like heart urchins and argonauts. We took these organic instruments to the percussionist Ronan Skillen, who became a kind of mentor to Tom, inviting him to his studio and teaching him to play various instruments from his collection.

The Voice of the Seaforest

Our song-catching project would truly find its voice when Ronan introduced me to the South African vocalist and songwriter Zolani Mahola. Zolani, who grew up in a land-locked township, loved the ocean, despite the systemic racism that had kept her locked out of the sea and virtually barred from nature. During South African Apartheid, beaches, like everywhere else, were strictly and violently segregated, and the few beaches where people of color were allowed were often inaccessible and dangerous.[4] Her family's daily struggle for survival didn't allow for many luxuries, and so only once a year did Zolani get the chance to be in her Mother Ocean.

I couldn't wait to take Zolani diving, and I was thrilled to discover she was one of those rare people who quickly adapt to the cold with little trouble. As we waded into the water, a school of tiny fish surrounded us.

She showed not the slightest fear of diving for the first time in cold water with no wetsuit. I could even hear her catching melodies: lilting Xhosa words that danced across the surface of the kelp forest.

I took Zolani on several diving and caving trips, teaching her the same way I had taught Tom: by letting her hold on to my back as we dove down deep. It took only two of these dives before she was free diving, climbing down the kelp stipes, gravity-free in the Seaforest. A land creature by birth, her soul craved the gravity freedom of her ocean heart.

"I'm cold drunk," she said, laughing, after one of our dives.

With Yo-Yo's visit fast approaching, Zolani and I began to work on lyrics. I would "catch" raw sentences and poetry during my dives—phrases like "free my amphibious soul" and "dreaming in the forest, forest in the dreaming"—and she would translate them to song in English and Xhosa. She was far better at songwriting and song catching than I was, but she was so kind and cooperative and somehow made my clumsy catching sound good.

Meanwhile, we began to gather the rest of our ensemble. Ronan brought aboard Jonny Blundell, a South African music producer who developed an intricate microphone setup to amplify and record the musicians, and acted as the creative glue between all the band members. Jonny in turn introduced us to his friend Madosini, a South African musical treasure who played Xhosa traditional instruments, such as the *mhrubhe*, or mouth bow. We were also joined by Pedro Espi-Sanchis, a Capetonian born in Spain who coaxed haunting melodies from the lekgodilo, a flute made from a length of dried kelp stipe. The combination of raw sounds from the wild and the ancient idea of song catching sparked a wildfire in everyone's minds.

A Longing for the Old Way

When the big day arrived, I felt an immense presence in the rooms of our beloved cottage by the sea. Though we had originally planned to have the concert outside, the threat of a storm forced us to improvise. I decided to host the gathering of about fifty people at our home. Swati was in India at the time, but Pippa and

Jannes were both patient with me, as I couldn't help but display some of my signature intensity. While not as bad as the terrible feeling of the past, the specter of my manic muse watched from the rafters as I raced about, making sure the artists had what they needed and that all our guests were comfortable.

After everyone gathered, the seated musicians began to play. I was so proud, watching Tom drumming alongside these seasoned performers, and though I was feeling the adrenaline rush of many weeks of preparation, as soon as I heard the sounds of the Seaforest, I began to relax. It felt like I was diving in a Seaforest where the water had been replaced by sound, thick and rich. My mind fell on the sound like a hungry person devouring a meal.

Then, in the midst of playing, Zolani paused, motioned for the band to stop. Something was wrong with the rhythm. I hadn't heard the mistake she'd heard, and for a moment my heart lurched with anxiety. But then she started again. The sound of Zolani's voice and the organic instruments my friends and I had worked together to create made me tremble in the way I often do when encountering profound beauty in the Seaforest.

Go down below the water
Oh, great love
Deep forest dreaming
You of flesh and blood
Hurinin, flow with me
Leave my head on the shore
Free my amphibious soul

Later, she would tell me that at one point in the song she had felt taken by something she didn't quite understand. In place of the words she'd planned to sing, a deep wailing poured from her throat.

It was haunting and powerful, a longing for the old way.

The ocean is the last wilderness on this planet. Zolani, like so many of us, craved full-body immersion in liquid bliss. In her voice, I could hear first the feelings of longing, then the sadness of the severing of our species from our Mother—the great forgetting—and finally, our reunion with the ocean.

YO-YO LOVED THE SONG CATCHING, AND AFTER THE SEAFOREST band finished, he played four cello pieces as a way to show his appreciation. Inspired by the meeting of nature and culture, he would become the patron of our Sea Change Project, and a wonderful supporter of our ocean conservation work.

Long after the concert, Zolani continued to dive and catch songs in the ocean. When we recently sat down to talk about her experiences, she told me she often saw images she couldn't explain during and after her dives—fractal shapes akin to rays of light. She began to have the distinct sense that someone was speaking to, or perhaps through, her.

Eventually she came to a profound realization: She was communicating with her ancestors. They were singing to her, and she was singing back.

As a girl, Zolani received a Western Catholic education, though her father also introduced her to traditional Xhosa ceremonies and healers. She told me that initially she found Xhosa practices, which are largely steeped in connecting to the ancestors, hard to grasp. It was another realm with so much to understand.

Now she began to leave offerings to her ancestors, feeling her connection to her Xhosa heritage deepen with every song caught.

"I almost feel like there are echoes of my great-grandmothers walking in the Seaforest," she told me. "And the more I speak my heart into the sea, the more I feel my heart open and grow . . . These messages I get from my ancestors light up my path. I can't go to the sea without it being sacred."

The song of the Seaforest is ancient and powerful. It carries the depth of every feeling humanity has ever known—our hunger, our longing, our sadness, our joy. It whispers to us about the wildness we've lost and urges us to search for it in places that can, at first, seem unfamiliar and cold.

But always, it dances with light.

"What a journey!" Zolani exclaimed, in a funny little exaggerated voice, after she finished sharing about the experience. Then she laughed her incredible laugh.

THE HEALING WILD

ONE MORNING IN MID-OCTOBER, WHEN SWATI WAS AWAY IN INDIA and Tom's friend Ben was spending the night on the first floor, I awoke early and started preparing breakfast in the kitchen. As I finished up, I heard a strange popping sound coming from upstairs where our bedroom was located and went to investigate. Halfway up the stairs, I smelled smoke and realized there was a fire in the bedroom.

At first, I thought I would just quickly put it out, but when I got to the bedroom door, it was already too late, flames were licking at the ceiling. In seconds, the room was an inferno, a giant roaring creature, and I knew we were in serious trouble. I ran downstairs to Tom's bedroom on the first floor and shouted at the two boys to wake up and get out. Together the three of us ran out of the house.

In moments, the upper story was engulfed, the fire already so hot that the windows were exploding and shards of glass were raining down around us as we ran. As soon as we were a safe distance away, I heard one of our neighbors shout that he'd called the fire department. And suddenly I remembered the propane gas cylinders—if they exploded, the neighbors' houses would go too; the whole neighborhood could burn to the ground. I thought back to the fire years earlier that had nearly wiped out the entire neighborhood, and knew we had to jump into action.

Tom and I slipped around the right side of the house, where the fire hadn't yet reached, and with shaking hands we desperately tried to unscrew the cylinders as the fire crackled and columns of black smoke billowed around us. We managed to uncouple the cylinders, but they were too heavy to lift, so we started dragging them away from the house. Finally, a neighbor came and helped us pull them away from the fire.

As soon as the cylinders were clear of the danger, we were able to take a breath. Tom and Ben and I just stood there huddled together in the blistering heat as the firefighters aimed streams of water at the fire, which roared and snapped like a living thing. We had taken nothing with us; there hadn't been time. We had just the clothing we were wearing and my cell phone. I stood and watched the flames consume our home—our dream house that Swati and I had found, that we had nurtured over many years, our beautiful house that held so many precious memories and artifacts that I had collected from all over Africa.

As the reality started to sink in, I noticed Tom looking at me as if he was trying to decide whether he should panic or stay calm. I realized I needed to make the same decision. I let out a deep breath and looked around.

"There's nothing we can do," I said to him. "Nothing at all."

And Tom stayed absolutely calm. He helped me get in touch with our insurance broker, then helped the firefighters put out the hot spots as the fire raged on for hours. Even after the blaze had been extinguished, the smoldering debris was so hot that embers kept flaring up, so we had to watch it all through that night and the next. I didn't sleep for two days.

On the day of the fire I had the terrible task of tracking Swati down in India to tell her what had happened. We had a bad connection, and she had trouble understanding what I was saying.

"I'm sorry, Swati. Our house is gone."

"What? What are you talking about?" It took her a long time to figure it out. She jumped on a plane and came back the next day.

ALMOST IMMEDIATELY, OUR FRIENDS STARTED TO SHOW UP. AT THAT point, I owned a pair of shorts and a T-shirt. Nothing else. No passport. No driving license. My friends Toren and Angus, who are my size, each gave me half their wardrobes—just gave me their own clothes. As word spread, dozens of other friends offered places to stay as long as we needed. A neighbor and friend Clive offered his garage to store anything we were able to salvage from the fire.

Such incredible kindness. Kindness and care and love from friends, neighbors, strangers, family members like my brother and

Lauren—people just helping us at every turn, guiding us through challenging decisions, like how to demolish, how to rebuild, and who would do what. All around me, guardian angels appeared, like my friend Sean, a master architect, who said, "I'm going to take care of this whole thing and you can worry about paying me later."

Our whole Sea Change team was a huge support, with Carina and Pippa going into overdrive to find help. Carina arranged for Guy and Dirk, experts on fire insurance, to help us. These two guardians battled the insurance company for nine months on our behalf, hardly asking for compensation. Craig Marais was a great friend, helping me carry remains of our house to safety with his vehicle.

I felt the deep kindness that is embedded in the human spirit all around me, which often appears most strongly in adversity. Swati's uncle and aunt, Prannoy and Radhika, gave up their house nearby and insisted we stay for as long as necessary. Calls started coming in from all over the world, words of love and kindness. My friends Al and Chris, who had dedicated much of their lives to shark conservation, called from a remote island in the Seychelles. I was so moved by their kindness and care, I shed a few tears. They said, "When it's all settled down, come visit our research station."

As soon as we got the fire mostly under control that first day, I knew one thing deep in my bones: I had to go into the water. Our home was destroyed, everything was total chaos, and after Tom drove away to stay with his girlfriend, Gen, I decided to walk down to the ocean with Pippa and Craig. My senses were greatly heightened from so much adrenaline. Big Mother lay before me—she really felt like my mother, her giant shiny surface coming toward me in the last light.

I was covered in soot, and it felt so good to immerse myself in the cool water. A spot on top of my head where a falling ember had burned my scalp erupted with pain in the salt, but I didn't care. I just lay on her surface and felt her life force flowing around me. This was my true home, my spiritual shelter. Nobody could burn her down.

And just that simple act of immersing myself in the sea soothed me. I was badly rattled and I knew recovering from the fire would be a massive amount of work and strain. But I also knew that as long as I had my health and my family and *this*—the ocean, nature, wild animals—I would be okay.

Salvage Work

Then began the long and drawn-out process of sifting through the ruins to try to recover what we could. The salvage was treacherous work. The upper story of the three-story house had burned and collapsed onto the two lower stories, and though one side of the house was completely gone, you could still inch up the staircase, which spanned gaps big enough to fall through. The wall supports were gone. A few rafters were hanging perilously, so we had to wear hard hats. As we crept carefully around, sifting through piles of soggy black debris, every now and again a joist or chunk of ceiling would fall and crash to the ground.

It was surreal to scarcely recognize the shell of what had been our home. We had made every little bit of that house picture-perfect, inside and out. I had painted everything by hand. Swati had always wanted a little glass-fronted steel fireplace near the kitchen and I had just completed this to surprise her homecoming.

Just as we'd truly brought every idea we'd had into being, it was all gone.

The fire inspector determined later that the cause of the fire was electrical, likely triggered by a power surge after a rolling blackout, a much-reviled practice in South Africa.

The firefighters had poured thousands of gallons of water over everything, and between the flames, the smoke, and the water, not much was left undamaged. The little natural swimming pond, once home to the Cape river frog, was coated with ash. My cameras and computers had melted and burned to a cinder, along with precious footage. Day after day, I sifted through the ashes like an archaeologist, sometimes with a shovel, sometimes with my hands, trying to find remnants of artifacts—artworks and musical instruments, shells and rocks, bones and seedpods, masks, notebooks, tapes and research: all irreplaceable, but still, all just objects.

There among the charred remains I found the handmade spear !Nqate had given me. The wooden shaft was completely incinerated, leaving just the metal spearhead and my memories of him.

Woven into the Wild Fabric

The house Swati's aunt and uncle had given us to stay in while we rebuilt was located near the penguin colony at Boulders Beach.

Getting to know the penguins was a bright spot during that dark time. But after days spent digging through the ashes of our home, and hour after hour on the phone with insurance companies, we realized that our souls and bodies desperately needed a break.

I had always fantasized about visiting a place that was virtually untouched, where animals outnumbered humans, where all the layers of species coexisted and I could feel the pulsing heart of the great Mother. I knew if I ever found such a place I'd feel my own heart beating the way it should. I'd be able to sense what my deep ancestors had sensed, feel the wholeness of being a human woven into the wild fabric.

THROUGH MY SHARK CONSERVATION FRIENDS, I FOUND IT IN THE outer Seychelles, a small group of coral islands and atolls whose near-pristine biodiversity offered a window into how our planet used to look. The area owes its rich biodiversity to its remote location and its proximity to very deep, nutrient-filled water, which is so nourishing to animal life.

This place was wilder even than the wildest places I'd been: the Okavango Delta, the Masai Mara, Central Africa, the Kalahari. Better yet, I was with my family: Swati, Tom, and Pippa, who was like a family member to us.

I awoke every morning at 5 a.m. to go tracking, and spent the day diving, documenting, eating and sleeping, taking in this paradise, drinking it in, letting the wild nutrients explode in my mind and fill me to the core.

The pulse of life here was remarkable. Every square inch was covered in some form of flourishing animal or plant life. Everywhere I looked, I saw turtles surfacing, squadrons of manta rays trawling for plankton, sharks of all sizes. When I dove off the edge of the shallow reef, large dogtooth tuna swam past me, giant shoals of fish swirled. I lost track of the days of the week—Monday to Sunday were replaced by manta ray day, shark day, turtle day, eel day, stingray day, fish day, and crab day. I could feel the island birthing so much life; this was how it was meant to be. I sensed the mind of a wild person infused with animal form, wildness dripping from every mind cranny, a body brimming with energy, each day a grand adventure. Most of all, I felt the healing powers of the island's immeasurable biodiversity.

Biodiversity—literally, the number of species of animals and plants—is the key to health, the engine of life on Earth. When we are in these rare places filled with huge numbers of animals, we feel joyous because of the beauty and life that surrounds us, but it is also restorative in a deeper sense—there is enough for all and we can relax, life is safe, our home is thriving, and there is no need to stress.

Nature in Charge

One day I traveled to a nearby atoll by boat.

I stepped off the boat onto a small uninhabited island. The first creature I saw was a red fiddler crab with one massive claw. The strange thing was the way the crab looked at me: I had the distinct feeling it had never seen a human being before. There was a kind of astonishment in its unblinking stare.

It felt so good to be in a place so wild. I entered the jungle, and amidst the foliage I made out the remains of several buildings, ruins from a bygone era. Trees had seeded in the cracks of the mortar and their roots were slowly breaking the building apart in ultra-slow motion. I was covered in mosquitos yet hardly noticed them in the wonder and intrigue of this place. This was one of those rare places on Earth where nature was overtaking human-made structures and thriving.

I saw massive golden orb-weaver spiderwebs, fifteen feet across, suspended between the broken buildings. Squadrons of land crabs scuttled under rusted old iron roofing, making an eerie but delicious sound. In the tea-warm shallows lapping against a ruined wall, several mangrove whiprays hunted for invertebrates. Tiny juvenile lemon sharks cruised among the rays. My mind feasted on the delicious sight of nature in charge.

I wandered around with my small camera, taking images of this lovely carnage, and climbed a set of steps leading nowhere. I had this vision of whole cities one day being consumed by the wild. Big Mother plays the long game as we scrabble to "win" the short set. She will teach us humility in the end. This vision was frightening and reassuring at the same time.

Being in this nature paradise and having daily interactions with so many species, seeing so many big fish and sharks, and spotting dolphins jumping out of the water in great numbers, had a profound effect on me that worked in two ways: it opened up my heart and made me fall in love with the wild all over again—and then it broke my heart to really feel how much we have lost.

Though it might take my family a few years to recover from the fire, we would surely have a home again. Our species, however, has no other home but Earth. How long will it take Earth to recover what she's lost? Her vanished wild species, her muzzled rivers and overheated oceans, her ancient trees and rugged mountains? How many years will it take to restore a melted glacier? A coral reef? A kelp forest?

I noticed a single impossibly long strand of spider silk undulating in the faint breeze that penetrated the jungle. A shaft of sunlight filled the silk with light, bright against the dark-green jungle background. I gazed down the thread as if looking down a telescope in time. I could feel the fragility and the strength of this single piece of spider silk.

The path was bright and shining—and then a cloud covered the sun and it was gone. But I didn't feel despair at its disappearance. I knew I would have to keep striving my whole life to keep that thread lit.

This was the world our species was born into, and this was the world our wild hearts were craving. My time on that island inspired me to think about ways we might regenerate our sacred ecosystems. What I'd learned from tracking in deep nature is that our most dangerous trait is our tendency to lose touch with our Mother and to think we can live without her. We need the wild for every breath we take, for every bite of food, for each bit of warmth.

My San mentors used to repeat things that they felt were important, and that's why I keep repeating this message: *What can each one of us do in our own small way to support biodiversity? What can each of us do to support the Mother of us all? She has worked twenty-four hours a day for almost fourteen billion years to get us here. She has fed us from the beginning; it's not much to ask that we now support her. All of our lives depend on it.*

Silver in the Dark

The next morning I woke early to track ghost crabs along the beach. I looked out at the crystal-clear water and then spotted something stirring about one hundred fifty feet out: a small black tornado swirling underwater.

When I waded closer I realized what it was—a bait ball, a last-ditch defensive measure smaller fish use to protect themselves when predators are nearby. I'd seen this phenomenon before in deeper waters but was amazed to see it happening in the shallows.

The tightly packed formation did not offer total protection. Frigate birds were dropping out of the sky to pluck fish from the churning water. Small blacktip sharks zoomed in the knee-deep water toward the tornado, their fins cutting the surface.

I rushed back to shore to grab my fins and mask and my little underwater camera, then dropped back into the warm water and swam toward the tornado.

Underwater, the battle for life was dramatic. Large predator fish moved like silver shadows, muscling into the tornado, gulping down the small fish. The sharks were a bit wary of me and kept their distance. Birds continued to spiral down from the sky.

By this time Tom and Pippa had joined me, and it was wonderful to watch this spectacle together. Despite how absorbing it was, I could still feel my attention pulled away to thoughts of the piles of work that awaited me on the other side of paradise.

Like hungry predators, fears and anxieties were eager to feast whenever they found an opening.

I changed my position slightly to photograph the tornado from another angle, and as the sunlight shifted, I saw that these fish were not black but silver. The tornado only appeared dark because there were so many fish and the ones highest up cast a shadow.

I felt the wings and feet of the birds skimming my back as they swooped low. And then that huge swarm, that tornado, enveloped me and I could feel thousands of tiny fish bodies moving against my skin.

I disappeared completely inside the tornado.

For a moment I wondered what was happening and then I realized these fish were operating as one tornado mind. They knew that I posed no threat—and actually served as a kind of barrier against the predators from the air and from the water. Neither the sharks nor the birds would attack in the presence of a large creature like myself.

In that moment the tornado in my mind went still. It was only the world around me and time rushing by that created the illusion of disconnection: predator and prey, animal and human, tame and wild.

I heard words in my head as if spoken by some old nature god. *It is silver in the dark—there is always silver in the dark.*

I felt still, quiet, there in the dark. *I am silver in the dark, I am silver in the dark,* and that's what I must keep telling myself when the adversity comes: *There is always silver in the dark.*

Then, as quickly as it began, the tornado dissipated. The smaller fish split up into groups; the predators glided in and scattered the fish. The sea was calm again.

I wondered what it was like to be a single fish in the tornado. For a brief fraction of a moment I had felt it, but before I could hold on to it, the memory was gone.

I can't remember that I was once a small fish. I can't remember that I was once a drop of ocean water. I can't remember that I was once an exploding galaxy. I can't remember that I'm made of the dust of ancient exploded stars. I can't remember that I'm more than twelve billion years old. I can't remember anything.

But for that instant I remembered who I was, and that's all I need. That distant memory, that tiny spark, that silver in the dark inside that swirling tornado—that's all I need, that spark to keep going, to keep living, to keep looking for the wild in this magnificent world.

HOW TO START YOUR OWN TRACKING PRACTICE

IT'S GOOD TO BE OUT IN NATURE JUST WALKING ALONG, BUT TRACK-ing can take you so much deeper.

For most of human history, tracking was a primal language that all humans could speak. As natural as breathing or walking, tracking is how we survived. Every wild child knew how to track, decoding the messages of each print and call, predicting what could be up ahead through animal alarm sounds and the smells carried by the wind.

Most of us have forgotten this language, but it's one we can relearn.

As with any language, first you learn the letters—with track-ing, that means getting to know individual species. A cricket, a sparrow, an otter, a snake.

Then you learn the words—the tracks and the most basic behaviors. Where does the cuttlefish live, how does it camouflage, what does it eat, how does it mate?

As you start to understand the more complex behaviors like threat postures, and basic predator-prey relationships, you get your first sentences going. Then, when you start to understand more deeply how this species interacts with that one, when you see how weather interacts with everything, when you get in sync with the rise and fall of the tides, or learn why the crickets like mating when it rains, that's when some of these sentences start joining together.

Now you are in conversation with nature. You can speak this wild language.

But where exactly to begin?

Tools of the Trade

Tracking does not require any fancy new equipment or special clothing; you probably have everything you need already. A small notebook to make sketches and keep notes will be useful. A good pocket tracking guidebook can help you identify species, or you can use online apps like iNaturalist. A ruler to measure the size of tracks and a pair of field binoculars may also come in handy.

One powerful tool that will enrich your tracking practice is a small pocket camera or high-end phone camera, and a mini tripod. The camera becomes an extension of your senses, because even most cell phones allow you to compress time in time-lapse or see the movements of animals in ultra-slow motion. You can also watch shy creatures while you're not present by setting up the camera remotely.

To keep the camera from getting in the way of the nature experience, I choose the smallest and easiest-to-use device that doesn't weigh me down or tire my senses with complex menus. The camera is a tool to help you understand the wild, take beautiful images, and relive your nature experiences while sharing them with others.

Over time you will build a wonderful collection of species, tracks, and behaviors, in images and notes—your own nature dictionary that will ultimately transform your relationship with your environment.

Don't rush. Enjoy this process of reconnecting with your wild self, with your design. Watch how it transforms your mind. The simple practice of looking for tracks and trying to solve nature's mysteries fulfills the wild human in us because it makes the mind work the way it was designed to.

Practice the Alphabet

The next time you go for a walk, take note of the animals you see and hear. Is there some bee or butterfly that draws your eye close? A swan floating in the reservoir on your bike route? A pair of finches that stops by your bird feeder at the same time each morning?

Pick species that appeal to you that you can engage with regularly. Tracking is most powerful when you can practice in your own backyard—whether that's the pond down the street, a nearby park, or a hiking trail. Your list doesn't have to be comprehensive; in fact, it's best to start with a group like insects or songbirds, which can be found everywhere. I've seen swooping owls and strutting peacocks in Delhi, a city grappling with an air pollution crisis. Otters swim in the canals of Cape Town. There are peregrine falcons living in New York City, nesting on skyscrapers and bridges and taking out pigeons in midair. In Central Park alone, there are more than two hundred species of birds![1]

Start with the animals nearest you, and once you learn one group—insects, for instance—you will also begin to understand how the lives of others work, and imagine what might be happening in their particular environment. You can then start figuring out things quite quickly, seeing patterns where before there was only puzzlement.

Don't shy away from developing a relationship with invertebrate animals. Jennifer Mather, a psychology professor at the University of Lethbridge and an expert on cephalopod behavior, recently shared with me some staggering statistics about the bias toward research on our fellow mammals, compared to insects, mollusks, or crustaceans. We know comparatively little about invertebrates, even though they make up 97 percent of the animals on this planet (most of them insects). She theorizes that this lack of knowledge may explain why so many people feel revulsion and fear toward our invertebrate kin.[2]

You may want to do your own research to create a list of some of the animals native to your area. Don't worry if there aren't a lot of animals around you. In a way, that makes it easier: fewer markings to decipher and remember makes tracking easier than in the African bush, where the maze of hundreds of overlapping tracks can be confusing. Take it slow and take note of just a few species on every outing. Think of it like collecting treasure. Once you start getting familiar with species, start to jot down further details about them. As you build your knowledge, you may wish to seek out like-minded people who have a specific interest in the animals or plants on your list. Compare species, check that your identifications are correct, and share observations.

Tracks Are Words

After the ABCs, you learn the words: the tracks and the most basic behaviors. Start familiarizing yourself with what tracks in your area look like. Initially, tracks may all look a bit similar, but as you get to know and recognize them, you'll see how very different they are.

After that, try to determine which direction the animal was going. How fresh is the track? Was the animal going slowly, dragging its hooves or paws? Or was it moving quickly and leaping about with lots of energy? If you are tracking on land, what do walking prints look like? What about running ones?

Tracking is not something you are going to learn overnight; it can take twenty years or more to master, but that doesn't matter. Just start small and chip away. Look for places where tracks are easy to find, like beaches and muddy areas. A fresh snowfall in colder regions can tell you a great deal about which animals share your hiking trails. Draw the tracks of animals you see. Look for signs everywhere, even around your home. Be alert for areas that are worn or scratched; seek out marks on plants, trees, rocks, or walls.

Tracks are not just marks on the ground but any clue left by any creature or plant, sand or rock. An animal's burrow, the marks from a bird's claws abrading a branch, a place where lightning has cracked a rock or scorched a tree trunk, or the marks left by running water can all be tracks. The grass moving in the wind leaves gorgeous concentric tracks in the sand and tells you which way the wind has been blowing. It tells you to keep upwind of animals you may be looking for, how the waves will break, and how the sand may sculpt the landscape.

Eventually you'll begin noting down animal behaviors and interactions with predators and prey. Start asking questions about every detail of the lives you are studying. Where does the animal live? How does it camouflage? What does it eat? How does it mate? What tracks and signs are associated with it? The more you know, the more interesting wild nature becomes.

If you see a turtle, ask yourself: *How does the shell grow? Does it shed? What is it made of? How does it remain shiny?* Make notes. Embed the wonders of nature in your memory. Sketch the process.

Consider bringing the different senses into your tracking practice. Though I personally have something of a mental block when it comes to birdcalls, a basic knowledge of the calls of ravens and kelp gulls has helped me track otters hunting in the Seaforest and the intertidal zone.

My friend Jon Young has mastered the art of bird language, which he shares in his fabulous book *What the Robin Knows*.[3]

Through years of close observation, Jon has realized that songbirds know everything important about their environment. By tuning in to their companion calls and warning alarms we can learn so much about the natural world around us, including what other animals may be close. Indigenous people worldwide have used this skill for millennia to survive and thrive.

Smell is another useful way to find animals and their predations. Jotting down descriptions of the important tracking smells helps me remember better—I find it a very useful way to identify which animals are nearby, often before I see them.

Tracking is a mindset; it's like becoming a detective in the show of life. When you first start to tune in to nature it can seem somewhat brutal. But learn its language and you start to see instead a deep, nurturing intelligence that is keeping the whole environment fertile, vibrant, and alive.

It's good to have a focus of curiosity and then follow the tracks to where that leads. If we are curious enough, and are prepared to spend time and energy observing the lives of wild creatures, they will allow us to step inside their secret worlds. I've discovered fish who can surf out of water to catch their prey, cephalopods who catch birds, fish that live partly upside down, and much more. All their lives are connected by thousands of interwoven threads of survival.

In Conversation with Nature

At the highest level of tracking mastery, you'll be following an animal in the wild by observing subtle signs and predicting unseen behavior and movement. But as a beginner, you may only be reading the language of the tracks left on the ground. You can go tracking and not see a single animal, yet you'll have a rewarding experience because you will see so much in the tracks.

Besides the physical tracks left by animals, you will also notice many other kinds of tracks: plants that have been nibbled on, tree bark that has been brushed up against, places where saliva or scat has been left. There are so many clues besides the marks animals leave on the ground. You can interpret these clues to put together a working hypothesis of what the animal might be doing.

Then, one day, you may see the behavior that answers your questions, proving or disproving your theories. It may be many months or even years after you first asked "Why?" But when it happens, you will feel a sense of profound wonder and gratitude that your tracking practice has helped you solve some wild mystery.

Connecting the Dots

Learning tracking takes persistence. Go out as often as you can. Consider scheduling a specific time each week, even if it's just an hour a week or twenty minutes each day. Regularity helps those random moving things become crystal clear until eventually you can see the patterns in the wild and even predict what animals will do. This is pattern recognition—something our brains are wired for because we've tracked from the beginning of human time.

Tracking demands—and rewards—your full attention. You're

not thinking of yourself or your fears and anxieties, or your job or the news of the day; you're just experiencing this beautiful flow as you work to connect the dots and solve a tracking mystery. Your mind and all of your senses are fully focused on the multidimensional world around you. The whole world becomes a magical place that you are in dialogue with.

It's an amazing feeling, similar to being completely absorbed by a sport or art, when you're lost in painting or sculpting, or in the groove of swinging a bat.

This ability to focus intensely on what's in front of you and what it's telling you is part of our ancient survival mechanism. Tracking is the key that opens that doorway to your old self. And when you feel that old self, you feel relaxed and confident, because there's no room for any other thoughts.

Afterward, you feel, *That was a day worth living.*

Ropes to God

The San's concept of "ropes to God" wholeheartedly captures the practice of tracking. Here is my interpretation of the process: You go outside in the morning and a bird lands in a nearby tree. You make a thread to that bird, just a gentle feeling of connection, acknowledging the bird's presence and its place in the world. Then you see a small insect climbing a branch. You make a thread to it. *I see you, insect. I'm grateful for what you do for our world.*

You do this with all the animals and plants you see, building these spider threads of connection and love. Then one day the threads weave together and form a rope, a conduit to the source of all life. The more detail and passion you put into making these threads, the stronger the rope will be. It's easy to feel adrift in a world that's essentially alien. It's easier to choose a path of wonder, as getting to know life's forces will help you leave despair behind.

We need relationships with wild creatures; we've had these threads running since our deep African beginnings. The insects and the wind are the timekeepers of the land tracker. The birds are our guidance system. The wild human mind is woven among the minds, sounds, and smells of wild creatures and places. Without these relationships we are bereft, only part human. The threads built between you and the smallest creature, plant, or stone lightens the soul, frees the mind. As I built relationships with wild creatures in the Seaforest and on the shore, I felt old pieces of my mind and body coming alive. I felt part of this place, connected and grounded.

As I write, I'm reminded of the time Jannes and I found a plank washed up on the beach covered in a colony of goose barnacles, a crustacean that attaches itself to rocks, driftwood, and other flotsam. We watched the barnacles feeding in the shallow water, and then found Columbus crabs living in the colony along with a pelagic nudibranch called *Fiona pinnata*, which feeds on the barnacles. Sadly, we knew none of these deep-sea animals would survive onshore.

We took a few of the Fiona eggs to the Sea Change Project's science lab and put them under the microscope. Magnified 180 times, this tiny egg mass the size of half my pinkie fingernail became a galaxy of pulsating life. Each living egg was like a crystal ball rotating on little vibrating hairs. This enchanted view built a strong thread to these nudibranchs that ride the high seas, sailing ships of flotsam. It was like stepping inside a new world, crossing a frontier. It made me realize that if I looked inside my own body, at my own living cells, I'd see the same thing, a wonder of biological splendor.

In that moment I built a thread to myself, to my own wild animal within.

Sitting Still

One valuable practice is to sit still for long periods to get to know nature in a certain spot. This is a tracking method I use to understand how a wild system works. Being still allows you to see small things and find clues, and animals will be far less afraid of a motionless human form.

Select a spot and try sitting quietly and just watching for an hour a day. My friend Jon Young calls this his "sit spot," a place where he goes each day and sits still and waits for creatures to get used to him and reveal themselves. It's a time to be quiet and observant; a kind of waking meditation, listening to the birds, slowly learning their language. It may not look like much, but actually you are observing a multitude of acoustic, olfactory, and visual signs.

Having a sit spot helps you develop a heightened understanding of your surroundings and slowly become a part of the natural environment. Your senses—sight, smell, hearing, touch, and even sometimes taste—are all valuable tracking tools, and working on improving them leads to greater awareness.

My sit spot is underwater in the kelp forest, where I go every day to learn and to feel alive. Choose a place close to where you live that is easy to get to quickly, so you can practice each day.

As a way to relax before dives, Jannes and I used to see how long we could lie still on the sand or grass before entering the Great African Seaforest. At first we got fidgety, and it wasn't easy to be still and just listen to the wavelets slapping and grunting against the rocks. Eventually we were able to be still for a long time, enjoying the rich flavor of the mind liquid bubbling to the surface, upwelling from the deep subconscious, filled with wild nutrients. That wild liquid fed lots of ideas—good, healthy ideas that would manifest in innovative science and storytelling, ideas like our 1,001 Seaforest Species project and our decision to include humans as one of those species. Wildness dies in the froth of cluttered, rushing life, but thrives in spaciousness. The wild mind needs space.

Building Bonds

Some people may enjoy tracking on their own, but most of us are social by design. We crave company, and we thrive from having close bonds with friends and family. Experiencing the wild with a friend or loved one is one of the greatest things a person can do. While solo time in the wild can be hugely beneficial, there needs to be a place of sanctuary to return to, and people who will gladly listen to your stories. Be careful not to use the wilderness to cut yourself off from society.

I have been fortunate to have many generous teachers and mentors on my tracking journey; you have met several of them in this book. I also cultivated deep friendships with some of the world's greatest archaeologists, Indigenous navigators, and other keepers of ancestral wisdom to learn more about the lives and practices of the first humans. I encourage you to identify people who can help you learn and to reach out to them with questions. Seek out mentors by writing emails or sending letters. Some people won't respond because they're too busy, but others will, and the result might be surprisingly gratifying.

Telling Stories

My San teacher !Nqate Xqamxebe was able to sense the presence of unseen animals in different parts of his body. He could feel the gemsbok, the lion, the springbok, the leopard, the porcupine, as an itching, as heat rippling through his legs and back and chest. Different parts of his body indicated to him which animals were close by in the landscape even when he couldn't see them. His body was a radar system tuned like a master instrument to the animals he loved and tracked. His flesh was made of their flesh.

Our deep ancestors had these gifts but without the trauma that !Nqate had to bear, the trauma of his people's genocide. The San have endured a litany of horrendous human rights abuses for several centuries and were even hunted with price tags up to the early 1900s. Through !Nqate I felt the tremendous power of the wild human spirit, of the sheer know-how of generations of wild people who'd come before, and I felt his deep pain too.

!Nqate died young, largely due to this pain, but his stories live on.

Storytelling is a key part of wildness and our most powerful tool for change. Our actions and emotions are shaped by our stories. For most of our time on this planet, our species did not write things down; instead, we passed on information by telling stories. These stories grew and died and were reborn many times, an enduring latticework of myth, legend, and survival knowledge.

The art of storytelling is embedded in our ancestral memory. Early humans, at first motivated by survival, took their art far past the need to survive, as their curiosity and wonder for natural science enlivened their giant brains. We crave story, and we remember stories hundreds of times more easily than facts. It's one of the reasons we love watching films so much, as they have taken this art form to an extraordinary level, involving hundreds of people coming together over long periods of time to tell a story.

After almost every good tracking session on land or in the water, I try to see the story nature has told me. I write down that story and draw on a lifetime in nature to enhance it, sometimes including a little research to add detail.

As you start your rewilding practice, try to get into a habit of finding the story that nature is telling you. Write down all you see and learn and then try to identify the story that most excites you. Then rewrite that story in the most fun and interesting way. Now you have something to share with friends or family, and the story will also allow you to remember what you have learned much more easily.

Observing animal behavior and finding exciting ways of re-telling their stories is perhaps the oldest practice on Earth. Our ancestors did it around night fires, which I love, but I mostly do it through writing and filmmaking. The wild book of love in my head finds ways to be shared with other people interested in the magnificent creatures of the Great African Seaforest.

A Universal Tracking Language

Is it possible to transfer tracking skills into a different environment? I'd been told that trackers outside their home territory would get lost and not be able to survive. But when I tested my tracking skills in the Seychelles, I discovered something quite different: that a universal tracking language can be applied to almost every ecosystem. I was able to apply the skills I'd honed over a decade in the Great African Seaforest to unlock the secret lives of eels and crabs relatively quickly when I visited these animals every day. It was exciting for me to start seeing under the surface of the nature matrix in that unfamiliar place.

When I'm tracking in wild nature there is a constant dialogue: the tracks are talking to me, and I'm talking back to them. I never feel lonely because I'm talking with nature the whole time, having this intense conversation, and then—*bang!* Sometimes there is this explosion as the tracks lead me to a moment of wonder, when nature shows herself. Those big aha moments—like finding a shell with a tiny drill hole left by an octopus—are tremendously exciting. Eventually you will be able to tell what animals were involved and what and how it happened, just by the tiny little mark left on the shell.

A Gentle Reminder

Remember that the animals you are tracking may consider you a predator, so keep your distance, as you may cause them stress if you come too close. Never trespass upon an animal's den, burrow, or nest; these are private homes that should be left untouched. If you find a baby animal, leave it alone unless you can confirm (by observing from afar for many hours) that it is orphaned and needs help. Many wild mothers leave their young alone for hours at a time while they forage for food, and they instruct their offspring to stay concealed in long grass or under a bush or log. If you or your dog come across one of these hidden babies while tracking, back away and leave it alone. Do not confuse stillness with calm; an immobile animal may actually be petrified, and some creatures can literally die of fright if handled by humans.

Making the Practice Your Own

A final word of encouragement:

The journey you are about to embark on is nothing short of life-changing. Opening yourself to the wild has a way of breaking down some of the barriers that humans have erected in order to survive in the strange technological world of our making. When those barriers crumble and the wild authentic person attempts to step out, it can be, at times, scary and even painful.

Perhaps the pain is there because old parts of the psyche have to die in order for you to awaken wild. Change is always frightening, even if it's positive. In moments when I find myself overwhelmed, I remember the basics: I breathe deeply, tap into the love I feel for my wild kin, or go for a walk or a swim with one of my fellow amphibious souls and open up about what I'm feeling.

Take it slowly and gently, and be kind to yourself.

It's my hope that you will adapt your tracking practice to suit your environment and your lifestyle. Make it simple and doable—an everyday practice that, if kept up over time, will nurture you. When you remember how to speak wild—your first language—you will find that it nourishes and invigorates you like nothing else. You are an integral part of the world around you; your life is an essential thread woven into your living planet.

ACKNOWLEDGMENTS

I've heard it said that writing a book is a lonely process, but I found it anything but. Between family, friends, mentors, and guides (both human and non-human), I felt supported and encouraged the entire time. I have deep gratitude for everyone who walked and swam with me on this journey:

My beloved octopus teacher, without whose presence in my life, this book would have never been written.

Rachel Neumann and Sarah Rainone, a dream literary team, whose vast experience in publishing, writing, and editing guided me every step of the way. By the end of the writing process, it felt like Sarah's and my mind were working as one. Special thanks to Tai Moses for her insightful edits, Hilary McClellen for her skillful research, and the whole team at Idea Architects, including Doug Abrams, Amanda Mikell, Lara Love Hardin, Bella Roberts, Janelle Julian, Mellisa Kim, Ty Love, and Ben Jahn.

Executive Editor Elizabeth Mitchell and President and Publisher Judith Curr at HarperOne are supportive visionaries who continually inspire me with their confidence. All their comments and advice lifted the book. My deep gratitude goes out to the talented and dedicated team at HarperOne, including Marketing VP and Deputy Publisher Laina Adler, Marketing Senior Director Aly Mostel, Publicity Senior Director Melinda Mullin, Art Director Stephen Brayda, Copyeditor Dianna Stirpe, Interior Designer Janet Evans-Scanlon, Design Manager Yvonne Chan, Managing Editor Suzanne Quist, Production Editor Lisa Zuniga, Assistant Editor Ghjulia Romiti, and Proofreader Theodore Kutt.

My wonderful friend Jane Goodall, PhD, DBE, founder of the Jane Goodall Institute and UN Messenger of Peace, for her ongoing encouragement to share my stories, and for enchanting me with her own powerful stories about nature.

My wife, Swati, my partner, collaborator, and greatest support system, whose fearless conservation efforts inspire me to deepen my relationships with my animal kin and whose smile lights up my life every day.

My parents, Keith and Diana Foster, for encouraging me to be independent from a young age; my grandmother and great-grandmother, Honey and Guggie, for listening so attentively as I told my first stories; my brother, Damon Foster, and sister-in-law, Lauren, my frequent adventure and filmmaking partners; my son, Tom Foster, with whom I continue my exploration of nature; and my cousins Sally Macey and Trish Neil for their ongoing support.

Sara Foster, my ex-wife, a wonderful co-parent and great friend, who has always supported my work. My fabulous young cousins Tara, Manya, Aidan, and Liam; John Macey, Paul Neil; and the Sundaram family.

ACKNOWLEDGMENTS

My extended family, including Krishna, Kannan, and Usha Thiyagarajan, whose deep understanding of philosophy and spirituality always leads to interesting discussions. Radhika and Prannoy Roy, who not only cheered me on but gifted me with cameras and kayaks to help support my daily dives. Brinda Karat, whose mere presence is like a shot of pure energy, and the wonderful Shonali Bose.

And where would I be without those friends who are close enough to be considered family? For years, I have benefited greatly from the love I've received from Jeremy and Guzzie De Kock, John and Karen Loubser and their daughter, Faine, and son, Tivon; Nirmala Nair, Dave Moore, Samantha McMurtrie, Marylin Macdowell, Bowen Boshier and Sally Andrew, Nick Ellenbogen, Lance Blaau, Yvette Oostehuizen, Lucretia Rodrigues, Gunter Pauli, Mike Kawitzky, Justine Mahoney, and Teo Biele.

I am so grateful to my Sea Change Project family, dedicated individuals working toward giving back at least a fraction of what nature has given us, including *My Octopus Teacher*'s codirector Pippa Ehrlich, a "young old soul" who is like a daughter to Swati and me; Carina Frankal, who became a lifelong friend after our adventures making a film called *Cosmic Africa*; Dr. Jannes Landschoff, whose brilliance and enthusiasm for wild nature taught me so much; Chris Van Mellenkamp, whose calm and wise presence helped us become a team; Daniel Ehrlich, whose enthusiasm and curiosity inspires me to tell stories in new ways; Levanah Grafton, who supports and nurtures at every turn; and the ever wise and kind Steven Frankal.

Deep thanks to all the generous souls who have supported the Sea Change Project over the years including the inspiring team at Plum Foundation—Sally Dufour, Germana Lavagna, Sandra Masato—who know the sea well after years of being close to her; Elizabeth Parker, who, with the support of the Mapula Family Trust, helped us stay afloat in the tough early years; Barend van der Vorm, who swam and tracked with us in winter; Liz Hosken of Gaia foundation; Zoe Vokes; Wim Hoff, Vivienne Boselak, Anya Adendorff, Christina Mittmeier of Sea Legacy; and Aaron Friedland, my frisbee mentor and friend. I would also like to acknowledge IPOS, the International Panel on Ocean Sustainability, especially Tanya Brody Rudolph and Francois Grail, for reaching out to Sea Change Project and involving us in the crucial work of ocean conservation globally.

Other people who have graciously supported Sea Change include Mike Nortje of Pisces Divers, Jonno Cope, Michael Daiber and the !Khwa ttu San Heritage Centre, master tracker Alex van der Heever, Ian Thomas, sculptor Robbie Rorich, Brent Stirton, Jerry Lemba Lemba, Colin Bell of Lekkerwater, Craig Fraser and Libby Doyle of Quivertree, Dominique Le Roux, Olympia Ammon, Nick Bezio, Dr. Adrian Nel and Dr. Amber Huff, Mike Kendrick and Harriet Nimmo, Matt Zylstra, Nik Rubinowitz, Scott Ramsay, Rob Ross, Sarah Waries, Gwen Sparks, Will Travis, Justin Blake, Ulrico Grech-Cumbo, and all the wonderful people at the Two Oceans Aquarium.

My life in the wild would not have been possible without the most extraordinary mentors, teachers, and guides who walked with me, shared their culture and knowledge, and taught me to track and understand nature and humanity's origins. My San mentors, including !Nqate Xqamxebe, Karoha Langwane, and Xhloase Xhokne, introduced me to the value of deep nature connection and primal joy, and my work with Xhosa master tracker JJ Minye continues to deepen my understanding of the first language. Professor Charles Griffiths, marine biologist and naturalist extraordinaire, shared his extensive scientific knowledge of the Great African Seaforest. Dr. Jeanette Deacon taught me so much about South Africa's extensive Rock Art sites. Professor Christopher Henshilwood enhanced my lifelong interest in the origins of humanity and, along with Dr. Karen Van Niekerk and Dr. Francesco D'errico and other scientists at SapienCe and WITS, was always there to answer my questions.

Jon Young and Anna Breytenbach further enhanced my deep connection with nature. Jurg Olsen and Karen Olsen have taught me so much about big cats. And Alwyn Myburg helped me explore wild Africa in a personal way.

My tracking partners Craig Marais, Gareth Fee, Kireon Mcshane, and Diogo Dominguez helped me enhance my ability to communicate with nature while deepening their own tracking skills.

Charmaine Joseph Gwaza, a Zulu sangoma (traditional healer) and my ex-partner, introduced me to her profound ancestral connection, which has continued to inform my work. Her insights and teachings have helped me deepen my understanding of my country, as have those of Lindiwe Dlamini, a sangoma whose family had deep roots in the freedom struggle, and Mbali Marais, a diviner in the Dagara tradition who has a deep passion for a return to Indigenous knowledge systems.

Nainoa Thompson, Hawaiian master navigator of the open ocean, has instilled within me a deep belief in the importance of sharing my tracking practice with the world.

This book would not have been possible without the contributions of a number of wonderful scientists who have always been there to answer my questions: octopus scientist Dr. Jennifer Mather; Petro Keene, an archaeologist and wonderful creative collaborator on our Origins exhibits; archaeologist Pieter Jolly; and anthropologist Dr. Megan Biesele.

I was also lucky to have the support of dedicated people like Cape Town's coastal manager Gregg Oelofse and councilor Aimee Kuhl, who believe in our work at Sea Change. The work of Professor Brian Swimme and Professor Louis Herman has expanded my mind considerably. I am so grateful to the unstoppable Dr. Sylvia Earle, who does so much ocean conservation and has been hugely supportive of our work at Sea Change. Professor George Branch, Dr. Ian McCallum, Dr. Peter Nilssen, Dr. Zach Bush, Dr. Renée Rust, Dr. Tony Cunningham, Professor Mark Gibbons and Professor Kerry Sink, Dr. Lauren de Vos, Dr. Dylan McGarry, and Dr. Tony Ribbink have all so kindly supported our work.

Pawan Patil and Anthony Mitchel, James Cameron, Maria Wilhelm, Dina Michelle, Pierre Morton, Ivy Givens, and Sy Montgomery all supported this book.

I have also received great support from the rangers and scientists at South African National Park, including Saskia Marlowe; shark scientist Alison Kock; Carl Notier of the marine anti-poaching unit; and the great Honorary Rangers team led by Keith McNair, Kenneth Carden, and Professor George Smith.

Ross Frylinck, who cofounded Sea Change Project with me and co-authored two books with me.

The De Hoop Collection, Nini Stephens and William Stephens, Dalfrenzo Laing and Hendrik Arendse, who have been so welcoming.

Dr. Dale Rae, Dr. Jamie Elkhorn, Andre Burger, Dr. Charles Chouler, Michelle van der Merwe, Dr. Murray Rushmere, Anja Gerbers, and Harold Epstein are all wonderful healers who I'm privileged to know.

In my long career as a filmmaker, I have been very lucky to work with excellent people, the best in the field, who made me a better storyteller. Ellen Windemuth, Ludo Dufour, Anne Druyan, Kent Gibson, Karen Meehan, Barry Donnelly, Kevin Smuts, Sophie Vartan, Kyle Stroebel, James Reed, Sarah Edelson, Jinx Godfrey, Roger Horrocks, Dan Beecham, John Chambers, Cristina Zenato, Tom Peschak, Walter Bernadis, Greg Thompson, Niobe Thompson, Micheal Raimondo, Warren Smart and Jackie Viviers, Helena Spring, Anant Singh, Nilesh Singh, Jason Boswell, Bryan Little and Fil Domingues, Donovan van der Hyden, Sam Barton-Humphreys, Neil and Nadine Clarke, and the late Michael Duffett.

A wonderful collaboration with Zolani Mahola, Yo-Yo Ma, Ronan Skillen, and Johnny Blundell resulted in a Seaforest Anthem, with Zolani becoming our Sea Change Ambassador and Yo-Yo our Patron, making possible another creative avenue for expressing deep nature connection.

Other collaborations like our 1001 Seaforest Species project led to a partnership with Save Our Seas Foundation and the great support of their CEO James Lea, a shark scientist, and former CEO Dr. Chris Clarke. They have since become dear friends, together with the inimitable Alma Artiaga, Stevo Morgan, and their dive crew, a friendship that has deeply enhanced our lives.

I am so grateful to His Excellency Abdulmohsen Abdulmalik Al-Sheikh, founder of the Save Our Seas Foundation, who has dedicated so much of his time and resources to saving sharks and rays around the world.

While writing this book, my house burned down, and what could have been intensely stressful was made manageable thanks to the help of special friends Angus McIntosh and Mariota Enthoven, Toren Wing, Cindy and Stuart Douglas; neighbors Rolf Sieboldt-Berry, Clive Stewart, Andre Van der Spuy, Gunnar Oberholzer. Extraordinary architect Sean Mahoney stepped in for the second time to save the house. Firemaster Greg Birch gifted his time; Chris Kisweter and Terrence McShane are handling the rebuild, and we couldn't have done anything without Guy Lloyd Roberts's and Dirk Kotze's help dealing with the insurance.

I would also like to thank Hanneli Rupert, present owner of my childhood home, who welcomed me so warmly into her home.

One of my San mentors, !Nqate Xqamxebe, spoke of the importance of "carrying a burning coal in the heart" from one place to the next, and allowing that coal to spark into life when needed. It's my hope that this book will act as a kind of kindling, lighting the ancestral fire within, reminding us of where we come from, and who we are.

I am so grateful to all the individuals who are helping to keep this fire burning.

NOTES

Chapter Two: Cold

1. In her book *Once Upon a Time Is Now* (New York: Berghahn Books: 2023), anthropologist Megan Biesele shares a pronunciation key for the special characters in the San language:

    / = dental click (cf. expression of irritation in English: "tsk, tsk").

    = = alveolar (laminal) click, no equivalent in English.

    ‡ = alveo-palatal click (cf. sound of a cork coming out of a bottle).

    // = lateral click (cf. sound used to urge on a horse).

2. Dr. Andrew Huberman, "Using Deliberate Cold Exposure for Health and Performance," *Huberman Lab*, podcast, 2:15:09, April 4, 2022, https://huber manlab.com/using-deliberate-cold-exposure-for-health-and-performance/.

3. Author interview with Wim Hof, October 18, 2021.

4. Wim Hof Method website, "Cold Therapy," https://www.wimhofmethod .com/cold-therapy.

5. Natalie Muller, "Eden's Killer Whales: Helping Human Hunters," *Australian Geographic*, October 12, 2012, https://www.australiangeographic.com .au/topics/history-culture/2012/10/edens-killer-whales-helping-human -hunters/.

6. Darren Incorvaia, "People and Animals Sometimes Team Up to Hunt for Food," Science News Explores, April 27, 2023, https://www.snexplores.org /article/people-and-animals-sometimes-team-up-to-hunt-for-food.

7. Francesca Trianni, "Otters Have Helped Bangladesh Fishermen Catch

Fish for Centuries," *TIME*, March 27, 2014, https://time.com/40632/bangladesh-otters-fishermen/.

Chapter Three: Track

1. Katherine Harmen, "Polarized Display Sheds Light on Octopus and Cuttlefish Vision and Camouflage," *Octopus Chronicles* (blog), *Scientific American*, February 20, 2012, https://blogs.scientificamerican.com/octopus-chronicles/polarized-display-sheds-light-on-octopus-and-cuttlefish-vision-and-camouflage/.

2. Author interview with Nainoa Thompson, November 2015.

3. Radio Expeditions, "An Interview with Anthropologist Wade Davis," by Alex Chadwick, NPR and National Geographic Society, May 2003, https://legacy.npr.org/programs/re/archivesdate/2003/may/mali/davisinterview.html.

4. Lyall Watson, *Lightning Bird: The Story of One Man's Journey Into Africa's Past* (New York: Simon & Schuster, 1982).

5. Author interview with James Lea, June 2023.

Chapter Four: Love

1. Author correspondence/interview with Janette Deacon, 2004.

2. Lawrence George Green, *Karoo* (Cape Town: H. Timmins, 1955).

3. Tony Jackman, "The Springbok Migrations," Africa Wild, africawild-forum.com/viewtopic.php?t=11477.

4. Dr. Jane Goodall, "Jane Goodall's Dog Blog—Rusty," *Best Pet* (blog), Perfect Pets, December 28, 2016, https://perfectpets.com.au/best-pet-blog/post/jane-goodall-s-dog-blog-rusty.

Chapter Five: Ancestry

1. Jason Daley, "Were Neanderthals Getting Surfer's Ear From Diving for Seafood?," *Smithsonian Magazine*, August 15, 2019, https://www.smithsonianmag.com/smart-news/neanderthals-had-lots-surfers-ear-suggesting-they-were-seafood-180972917/.

2. Author interviews with Christopher Henshilwood, 2013–2023.

3. Ed Yong, "An Ancient Crosshatch May Be the Earliest Drawing Ever Found," *The Atlantic*, Sept. 12, 2018, theatlantic.com/science/archive/2018/09/is-this-the-earliest-drawing-ever-found/570007/.

4. Brian Thomas Swimme, *Cosmogenesis* (Berkeley, CA: Counterpoint, 2022).

5. Jane Goodall Institute, "Jane Discovers That Chimpanzees Make and Use Tools," https://janegoodall.org/our-story/timeline/.

6. Margaret Roberts and Sandy Roberts, *Indigenous Healing Plants* (Johannesburg: Southern Book Publishers, 1990).

7. Author personal communication with Tony Cunningham, 2015–2022.

Chapter Seven: Connect

1. Scott Neuman, "Revenge of the Killer Whales? Recent Boat Attacks Might Be Driven by Trauma," NPR, June 13, 2023, https://www.npr.org/2023/06/13/1181693759/orcas-killer-whales-boat-attacks; and Stephanie Sy and Courtney Norris, "Group of Orcas Attack and Sink Vessels off Iberian Coast," *PBS NewsHour*, video, 4:25, June 14, 2023, https://www.pbs.org/newshour/show/group-of-orcas-attack-and-sink-vessels-off-iberian-peninsula.

2. Phoebe Weston, "Orcas Accused of Attacking Boats May Be 'Following Fad' Scientists Say," *Guardian*, August 25, 2023, https://www.theguardian.com/environment/2023/aug/25/orcas-boats-rammings-scientists-open-letter-aoe.

3. James Fair, "Are Killer Whales Dangerous to Humans?" *Discover Wildlife*, BBC Wildlife, January 24, 2023, https://www.discoverwildlife.com/animal-facts/marine-animals/are-killer-whales-dangerous-to-humans.

4. Sophia Ankel, "Hundreds of Great White Sharks Have Vanished from South Africa's Coast and Fearsome Orcas Are to Blame," Insider, November 22, 2020, https://www.insider.com/orca-attacks-caused-great-white-sharks-flee-cape-town-experts-2020-11.

5. Boris Worm, "A Most Unusual (Super) Predator," *Science* 349, no. 6250 (2015): 784–85, https://www.science.org/doi/10.1126/science.aac8697.

6. Sylvia A. Earle (@SylviaEarle), "Sharks are beautiful animals," Twitter (X), July 16, 2021, 5:07 p.m., https://twitter.com/SylviaEarle/status/1416187797963116545.

7. Thais Martins, et al., "Intensive Commercialization of Endangered Sharks and Rays (Elasmobranchii) Along the Coastal Amazon as Revealed by DNA Barcode," *Frontiers in Marine Science*, December 14, 2021, https://www.frontiersin.org/articles/10.3389/fmars.2021.769908/full.

8. Katherine J. Latham, "Human Health and the Neolithic Revolution: An Overview of Impacts of the Agricultural Transition on Oral Health, Epidemiology, and the Human Body," *Nebraska Anthropologist* 28 (2013): 95–102, https://digitalcommons.unl.edu/nebanthro/187/.

9. Henry Kam Kah, "The Laimbwe Ih'neem Ritual/Ceremony, Food Crisis, and Sustainability in Cameroon," *Journal of Global Initiatives* (2016): 53–70, https://www.academia.edu/23789087/The_Laimbwe_Ihneem_Ritual _Ceremony_Food_Crisis_and_Sustainability_in_Cameroon.

10. Library of Congress website, "America at Work," loc.gov/collections /america-at-work-and-leisure-1894-to-1915/articles-and-essays/america -at-leisure/.

11. Author personal communication with Nainoa Thompson, November 2015.

12. Nainoa Thompson, "As a Wayfinder, as a Native Hawaiian, and—in the End—as Nainoa, What Is Sacred?" in *Sacred: In Search of Meaning,* by Chris Rainier (San Rafael, CA: Mandala, 2022), 258.

13. oceanexplorer.noaa.gov/facts/pacific-size.html.

14. The Digital Bleek and Lloyd Collection, lloydbleekcollection.cs.uct.ac.za /index.html.

15. lloydbleekcollection.cs.uct.ac.za/index.html.

16. Author interview with Renée Rust, June 2020.

17. The Mother Tree Project, mothertreeproject.org/about-mother-trees-in -the-forest/.

18. Author personal correspondence with Jannes Landschoff, et al., 2023.

19. Jessica Wimmer and William Martin, "Likely Energy Source Behind First Life on Earth Found 'Hiding in Plain Sight,'" Frontiers, January 19, 2022, https://blog.frontiersin.org/2022/01/19/frontiers-microbiology-origin-of -life-energy-hydrothermal-vents/.

20. Brian Thomas Swimme, *Cosmogenesis* (Berkeley, CA: Counterpoint, 2022).

21. "How Human Beings Almost Vanished from the Earth in 70,000 B.C.," NPR, October 22, 2012, https://www.npr.org/sections/krulwich /2012/10/22/163397584/how-human-beings-almost-vanished-from -earth-in-70-000-b-c.

22. Rutger Bregman, *Humankind: A Hopeful Story,* trans. Elizabeth Manton and Erica Moore (Boston: Little, Brown, 2020).

Chapter Eight: Play

1. Ben Garrod, "Can All Primates Swim?" *Discover Wildlife*, BBC Wildlife, May 25, 2023, https://www.discoverwildlife.com/animal-facts/can-all -primates-swim/.

2. Bluegrass Music Hall of Fame and Museum website, "Bill Monroe," www .bluegrasshall.org/inductees/bill-monroe/#biography.

3. Anthony D. Fredericks, "How Engaging with Nature Bolsters Creativity in Children and Adults," *Psychology Today*, July 7, 2021, https://www.psychol ogytoday.com/us/blog/creative-insights/202107/how-engaging-nature -bolsters-creativity-in-children-and-adults.

4. Hanibal Goitom, "On This Day: Desegregation of South African Beaches," *In Custodia Legis* (blog), Library of Congress, November 16, 2015, https ://blogs.loc.gov/law/2015/11/on-this-day-desegregation-of-south-african -beaches/.

Learning the Wild Language: How to Start Your Own Tracking Practice

1. Meghan Bartels, "The Insider's Guide to Birding in Central Park, New York City," *Audubon Magazine*, June 2, 2017, https://www.audubon.org/news /the-insiders-guide-birding-central-park-new-york-city.

2. Jennifer A. Mather, "Ethics and Invertebrates: The Problem Is Us," *Animals* 13, no. 18 (2023): 2827, https://www.mdpi.com/2076-2615/13/18/2827.

3. Jon Young, *What the Robin Knows: How Birds Reveal the Secrets of the Natural World* (New York: Houghton Mifflin Harcourt, 2012).

A NOTE ON THE COVER

THE BLUE BUTTON, PORPITA PORPITA, IS THE ANIMAL DEPICTED ON the book's cover. It's an open ocean free drifter, sometimes ending its journey in the Great African Seaforest. This creature is a member of the family Porpitidae and has two main body parts: a brownish center known as a float, which is surrounded by a colony of hydroids that resemble jellyfish tentacles. It has both male and female organs, and can reproduce on its own. Its engines are the tide, wind, and swell. To me this creature looks like a planet made mostly of ocean with a small landmass—blue, wild and free.